CAMBRIDGE COUNTY GEOGRAPHIES

General Editor: F. H. H. Guillemard, M.A., M.D.

# NORTHUMBERLAND

*Cambridge County Geographies*

# NORTHUMBERLAND

by

S. RENNIE HASELHURST
M.Sc., F.G.S.

With Maps, Diagrams and Illustrations

Cambridge:
at the University Press
1913

CAMBRIDGE UNIVERSITY PRESS
Cambridge, New York, Melbourne, Madrid, Cape Town,
Singapore, São Paulo, Delhi, Mexico City

Cambridge University Press
The Edinburgh Building, Cambridge CB2 8RU, UK

Published in the United States of America by Cambridge University Press, New York

www.cambridge.org
Information on this title: www.cambridge.org/9781107677500



First published 1913
First paperback edition 2013

*A catalogue record for this publication is available from the British Library*

ISBN 978-1-107-67750-0 Paperback

# PREFACE

IN writing this volume I have had to refer frequently to original records and researches. Particularly must I mention the assistance obtained from the *Transactions* of the Natural History Society of Northumberland and Durham, Bruce's *Roman Wall*, and Professor Lebour's *Geology of Northumberland and Durham*.

My sincere thanks are due to several friends for kindly help and generous criticism; especially to Mr E. W. Heaton, B.Sc., Headmaster of the Tynemouth Municipal High School, who has kindly read the MS.; as well as to Mr R. E. Rigg, B.A., Mr George Earnshaw, and Mr Alfred Hair, Librarian to the Borough of Tynemouth, who have greatly helped me with certain sections.

I have also gratefully to acknowledge the aid of Professor Meek, Mr McDonald Manson of the Tyne Improvement Commission, and Mr Baldwin of the Blyth Harbour Commission, to whom I am indebted for most useful statistics, while to Dr Guillemard, General Editor of this Series, Mr F. W. Shields, M.A., and Mr W. H. Young, F.L.S., I tender my thanks for valuable criticism.

S. R. H.

NORTH SHIELDS.
*November*, 1913.

# CONTENTS

# ILLUSTRATIONS

## MAPS



The illustrations on pp. 5, 7, 9, 12, 21, 32, 34, 39, 41, 45, 51, 72, 96, 99, 101, 117, 125, 127, 129, 134, 136, 137, 139, 141, 146, 148, 163, 167, 172, 173, and 174 are from photographs by Mr Godfrey Hastings, of Newcastle; those on pp. 17, 18, 19, 22, 48, 97, 104, 110, 112, 113, 118, 119, 124, 156, and 169 from photographs by Mr J. P. Gibson, of Hexham; those on pp. 30, 42, 74, 78, 81, 92, and 177 from photographs by Mr W. H. Elliott, of North Shields; those on pp. 126 and 133 from photographs by Mr G. W. Wilson; that on p. 135 from a photograph by Messrs J. Valentine and Sons. The sketch-maps and plans are from drawings by the author.

# 1. County and Shire.

Scarcely anything indicates the age of a country or county so well as the manner in which its frontiers are delimited. A glance at maps of North America and Europe shows this at once. In the former the boundaries of the states and provinces are mathematically precise and are not dependent in the main upon physical features, and in the latter we find that there is an interdependence between frontier and natural feature not only as regards the major states, but even in the smallest counties. The expansion of power of two rivals on opposite sides of a natural frontier generally resulted in the adjustment of the frontier along ridge and river. This is particularly noticeable in North England and especially in the case of Northumberland.

The words county and shire are commonly used to designate the main divisions of our country. County (French *Comté*) is a relic of the Norman period and was the territory of a Count (*Comte*) or Earl, but this term acquired a wider significance and is now applied for the purposes of financial and judicial administration. In this sense Sussex, Kent, and Essex are counties, but they are in reality only remnants of old Saxon kingdoms. In many cases the older Saxon appellation shire (A.S. *sciran* —to shear, to cut off, a word which implies that the parts were shorn off from major divisions) has been superseded by county. Both terms are still used; for instance we should use shire in speaking of Buckingham, and county in the case of Durham. The older shires of Saxon times had each to provide their portion of the King's revenue and supply a certain number of men-at-arms in times of war, and they were virtually governed by a shire-reeve (sheriff). At the Conquest the county displaced the shire.

With reference to the name of the County of Northumberland it is generally conceded that it was given by the Saxons, Northan-Hombra-lond. The name is almost self-explanatory—the land north of the Humber—but it is needful to say that in the vicissitudes of war the boundaries varied considerably and were always under adjustment. Northumberland originated out of two kingdoms, Bernicia and Deira, which coalesced in 593 under Ethelfrith. It was twice again severed, in 633 and 642, when the two kingdoms were again formed. Again re-united they remained so till 840, when Northumbria ceased to be independent and was subject to Egbert of Wessex. It was reduced to an Earldom by Edred, who made Osulf the first governor or earl in 945. This office continued till after the Norman Conquest, when a *vicecomes* (viscount) or High Sheriff superseded the Earl.

Northumberland extended from the Forth to the Humber. That portion between Edinburgh and the Tweed was the scene of the ravages of the Picts and

Scots. It was known as the Lothian, and included Berwickshire, Teviotdale, and East Roxburghshire. It was ceded to the Scots by Eadulf, Earl of Northumberland. It was, however, still looked upon as part of England, being peopled mainly by Saxons, and Malcolm IV paid homage for it to the English king. Nevertheless, Berwickshire still remained for long an object of dispute, though from this period onward the boundaries of the county were practically what they are to-day. Until the reign of Henry VIII the county of Northumberland included several detached portions of the county of Durham, i.e. Norhamshire, Islandshire, and Bedlingtonshire—a relic, no doubt, of the time when Northumberland was under the Bishop-princes of Durham (Palatii Comites).

## 2. General Characteristics — Position and Natural Conditions.

Northumberland in many ways combines the characteristics of a variety of counties. It is maritime, agricultural, and industrial. The long coast-line with its pleasant little bays, combined with the nearness of the fishing grounds, has caused a fishing community to spring up. In the larger harbours like the Tyne and Blyth the seafaring instinct is largely developed. But, after all, the county is primarily industrial. The vast coal supply has attracted a huge mining population, very distinct in type, language, and customs, and there has also arisen the concomitant engineering community. Northumberland possesses an extensive coastal plain and, as quite a large proportion of the county is below 300 feet, agriculture is fairly well developed. Cultivation extends up the arms of the valleys. Above 600 feet the land acquires the general aspect of typical moorland, given up in the main to raising hardy cattle and sheep. Thus a large number of the people are scattered, and engaged in rural occupations.

Generally speaking, true rocky peaks are absent. The hills which occur are either rounded or are scarped lands. Where these are dissected by streams we get more or less tabular hills. The term "crag," so common in the county, fittingly recalls the normal scenery. The slopes away from these crags are covered either with coarse moorland grass or heather, and are often boulder-strewn, with the bed-rock protruding. The rivers are well developed. In their upper reaches the gradient is steep, but usually they are meandering streams with pleasantly wooded slopes until they enter the industrial area, where the landscape is often dismal. Of lakes, most of which are little more than ponds, fourteen are called loughs. This is remarkable, for we find the term lake of universal use quite near in Cumberland. The area known as the "Northumberland Lakes" is undoubtedly lovely, with a wild majesty in its scenery. One of the finest features of the county is the number of *denes* or pretty wooded valleys with streamlets passing through them, among which Jesmond Dene is famous. Commons and forests occur above 600 feet, but these are

Jesmond Dene

but other words for moor, and most likely are relics of old deer-walks. Woods do occur, but here again the term is practically supplanted by dene or lynn.

The climate is generally rigorous and humid, despite the small rainfall. This is particularly noticeable in clayey areas. Scotch mists and fogs are common along the coast. The scenery, away from the industrial area, is always interesting but never grand, except in the wild parts of the Whin Sill, the Cheviots, or the South Tyne. Waterfalls are not common, and when found they occur in the upper reaches of the rivers where they pass over the hard strata. There is a distinctive similarity about Northumbrian landscapes, and the reason for this will be seen later when we consider the geology of the county.

## 3. Size. Shape. Boundaries.

Northumberland lies on the borders of England and Scotland, between Berwickshire and Roxburghshire on the west and north, Cumberland on the west and south, Durham on the south, and the North Sea on the east.

Its greatest length, measured from near Berwick to the border at Allenheads, is nearly 71 miles. Roughly quadrilateral in shape, it has an area of 1,291,530 acres, or roughly 2018 square miles. Relatively to the other counties of England it thus stands fifth in size, being of about the same area as Lancashire, larger than Somerset, but smaller than Yorkshire, Norfolk, Devonshire, or Lincolnshire. The boundaries are very simply defined by the prominent natural frontiers of the Cheviots and the northern termination of the Pennine axis. Northumberland is cleft by the Tyne Gap, in such a manner as to include most of the basin of that river in her area. A conspicuous feature of the demarcation of frontier is that it usually follows a river till it becomes fordable and

Cheviot Hills from Dod Law

then almost invariably turns to the nearest high land. This is well instanced in the Tyne, where it follows the river from Tynemouth to Wylam, nearly 18 miles, and also in the Tweed, between Lamberton and Carham—also nearly 18 miles. At this point the Tweed flows in a general north-easterly direction. At Carham the boundary turns almost through 90° south-eastwards for

14 miles, along high land, arriving thus at the Cheviot mass. From this point it passes along the Cheviot ridge for 28 miles south-west to Caplestone Fell. It then takes a general southerly direction following the river Irthing past Gilsland in a more or less sinuous course, and thence ascending the high moorlands of the Pennines to Whitley Common, a distance of nearly 33 miles. Between Humble Hill and Gilsland the Irthing is backed on the east side by a ridge of hills from Bolts Law through Black Fell, thus forming a double frontier. The south boundary passes almost east by north from just south of Allenheads *via* Blanchland to Shotley Bridge, a distance of 34 miles, keeping to the Derwent all the way from two miles north-east of Allenheads. Just beyond Ebchester it leaves the Derwent and strikes northward for the Tyne (being here on the fringe of the coastal plain) which it joins at Wylam, distant about eight miles. It now follows the Tyne to Tynemouth.

The seaboard of Northumberland roughly measures 70 miles from near Lamberton to the Tyne. Wherever hard rock like limestone, sandstone, or basalt reaches the coast it produces distinct cliffs of a very rugged and treacherous nature. Monotonous lengths of clay-cliff occur for miles in the intervening stretches, wherever glacial action has scoured away the bed rock or where it has been removed by other agency. Its many sand-dunes, drift carried by the stiff north-east gales, beaches, and gently recessed bays (as at Alnmouth, Amble, Druridge, Newbiggin, Seaton Sluice, Whitley, and Tynemouth) are attractive alike to naturalist and pleasure-seeker.

Of islands, Holy Island, the Farnes, Coquet Island, and St Mary's Island are conspicuous. The dangerous north-east gales make shipping perilous at times, though protection is afforded by the fine stone piers and break-waters erected along the coast. Many stately castles, abbeys, and ruins cap the headlands, notably those of Bamburgh, Dunstanburgh, and Tynemouth. The treach-erous nature of the Farne Islands and of the Black

The Farne Islands

Middens at Tynemouth is well known and is recalled by the names of Grace Darling the heroine of the Longstone, and Greathead the inventor of the lifeboat.

The Farnes are chiefly naked basaltic rocks—out-standing masses of the Great Whin Sill—about 25 in number, lying from one to five miles east from the coast. Black pinnacles of columnar basalt rise in parts 50 feet high, and are cleft by deep chasms. The Farne, the

Wedums, the Noxes, Staple Island, the Fosseland, the Wamses, the big and little Harcar, and the famous Longstone Ridge are the chief of the group. Lying to the north-west, and about six miles away from the Farne, is Lindisfarne or Holy Island. It is a mass of limestone, sandstone, and basalt, about two miles long from north to south and one-and-a-half broad, excluding the Snook or spit. Holy Island, whose scenery has been celebrated by Turner, possesses a remarkable history in the annals of Christianity. The forbidding nature of this coast is strikingly exemplified also at Crag Point and Dunstanburgh, and at Cullernose Point, where the basalt rises to a height of 120 feet.

A submerged forest occurs at Howick, having revealed (about 1849) remains of oak, hazel, and alder, many of the trees being *in situ*.

## 4. Surface and General Features.

The topography of Northumberland bears a simple relationship to its geology, for the high land coincides with the harder rocks and the low land with the softer. Thus we get a series of ridges, crags, edges, or scarps with a certain general linearity from south-west to north-east, and with their faces directed inland towards the Cheviot. Passing from the central high area to the south-east we enter cultivated tracts with "dry-diked" enclosures on the hill-sides, and arable lands and farms lower down.

The colour-changes of a Northumbrian landscape are often very rich, but the dominant colour is brown. This frequently conveys an impression of expanse and altitude which is in reality non-existent and may be attributed to the undulatory nature of the land in the fell country.

Northumberland may be conveniently divided into three main surface divisions, each of which bears a distinct relationship to the rocks—namely the Cheviot, the Crag-lands, and the Coastal Plain. The Cheviot includes the area extending from the Redewater to Wooler and Alnham. The Crag-lands extend as a band, with the last as a boundary, and include the whole of the Carboniferous beds to the top of the Millstone Grit. This may be further subdivided into two types, the sandstone and the limestone, according to the nature of the predominating rock. The southern and eastern limits of this area we can place along a line joining Amble, Morpeth, Ponteland, and Bywell. The Coastal Plain is taken to include the area east and south of the latter and is bounded by the coast. It coincides with the true coal-field. These districts we shall now consider in detail.

By far the most impressive is the Cheviot mass. The hills composing it are formed of a variety of igneous rocks like granites, porphyries, lavas, and tuffs. They are chiefly conical in shape, and present rounded green contours. They cover an area in Northumberland of about 200 square miles. The flanks of these hills are scored by deep glens, and in the distance one can scarcely mistake their origin, for the outlines suggest it. They culminate in Cheviot (2676 ft.); other peaks being Cairn Hill

(2545 ft.), Hedgehope (2348 ft.), Comb Fell (2132 ft.), Windy Gate (2034 ft.), Cushat Law (2020 ft.), the Schel (1979 ft.), Dunmore or Dunmoor Hill (1860 ft.), Black Hag (1790 ft.), Newton Tors (1762 ft.), and Hungry Law (1643 ft.). For several miles the average elevation is over 1800 feet.

The Cheviot Hills

Passing from the Cheviot in the direction of Alnham it is interesting to note that villages spring up practically at the entrance to the Crag-lands where the fast mountain torrents debouch upon a more or less flat area lying between the Cheviot and the Rothbury Forest. The Lower Carboniferous rocks, stretching around the *massif* of the Cheviot, are mainly sandy and extend from the Tweed through Harbottle to the Redewater. They contain limestones but of a different nature from the Fell-top and Great limestones, and the scenic type is also different. The general elevation of this series is from 600 feet upwards. The "limestone" type of the Crag-lands consists of a variety of rocks comprising limestones, grits, sandstones, shales, and even coal.

The "gritty" type is best exemplified in the Harbottle Grits forming Harbottle Hills, Harbottle Crag, and High Spoon Hill (1164 ft.). These grits are thickest in the Harbottle district at the Beacon (988 ft.), but swell out to form the wonderfully fine crag scenery of the area lying to the south-west between the Redewater and the North Tyne. They extend as far as Peel Fell (1975 ft.), Carter Fell (1815 ft.), and Deadwater Moor (1867 ft.), and frequently form great amphitheatres in the hills at an elevation of 1000 feet. This grit type of scenery disappears away to the north-east from a line joining Harbottle with Tosson Hill to the south-east of it, and this is only to be accounted for by the sudden change in the type of rock met with. The Rothbury Grits form the Simonside Hills of the Rothbury Forest, culminating in Tosson Hill (1447 ft.) and Simonside (1409 ft.). Long Crag in the North Forest is 1000 feet high. These grits form the characteristic scenery of Callaly Long Crag (1047 ft.) and extend to the Wishaw Pike (1000 ft.) and Hartside Crag near East Woodburn. The identical Crag-lands scenery is maintained across the North Tyne Valley in alignment with the Roman Wall and to the north side of it, forming the Bell Crags (1089 ft.) of Haughton Common.

To treat of the area south of the Tyne from Glendue Fell (1711 ft.) to Blanchland Moor (1345 ft.) would be but to repeat the description of the scenery of the Rothbury Grits, for this area is mainly "grit land" of Millstone Grit age. The chief topographical features are Pike Rigg (1723 ft.) on Whitfield Moor, and Hartley Moor (1876 ft.) on Allendale Common. The limestone type forms a general band of contour from 300 to 600 feet high. The Great Limestone affords a striking topographical feature and can be traced right across the county. It is best seen near Great Whittington and Stagshaw Bank on the Roman Wall. So varying is the nature of these limestones that we can only consider them as a distinct surface feature in the triangle formed by the Whin Sill, the river Tyne, and the main road from Newcastle to Belsay and Scots Gap. The Fell-top Limestone is best seen at Harlow Hill (505 ft.), and between it and the Little Limestone at St Oswald's Chapel the land rises to over 800 feet. The general trend of the beds is to the north and north-east.

Passing almost midway between the "Grits" and "Limestones" is the line of the Great Whin Sill. This igneous rock is a dominant factor in the scenery in many parts and merits special mention. It also belongs to the "Crag-lands" division. It passes into the county four miles west of Haltwhistle at Greenhead and keeps a north-east trend (being capped by the Roman Wall for many miles) to Low Teppermoor. At Hotbank Crags it is most impressive, standing 1080 feet above the sea. It strikes across the North Tyne for Gunnerton Crags and passes through Little Swinburn, Sweethope Burn, Knowes Gate, Gallows Hill, Greenleighton, and Ward's Hill, all of which are from 600 to 700 feet above the sea. It then disappears for several miles and is not met with again till we get to Rugley, south of Alnwick. It then strikes for the coast at Dunstanburgh, forming the high land at Peppermoor, and to it is due the fine cliff scenery at Dunstanburgh, Bamburgh, and the Farnes. From Bamburgh it runs inland and forms the Spindleston Crags, and then through Middleton to make the Kyloe Hills, south-east of Lowick.

The Coastal Plain has a general elevation up to 250 feet and forms an undulatory surface given up to arable lands and pasture. It is generally glacially covered and is the area of the true coalfield. Its low contour has greatly facilitated the development of roads and traffic lines. The Permian strata so characteristic of the county of Durham form a tiny patch of fine cliffs at Tynemouth over a hundred feet in height.

## 5. Watersheds. Rivers. Lakes.

Northumberland is a typical instance of a true natural geographical region. The chief watershed of Northumberland comprises the Cheviots and the northern termination of the Pennines. These extend in a horse-shoe divide from Wooler to Ebchester. Thus the general slope of the drainage of the county is to the south-east north of the Tyne, and to the north and north-east south of the Tyne. There is one prominent exception to this general water-parting, namely the Tweed-Till basin in North Northumberland—the Till flowing around the Cheviot foot-hills towards the north and north-west. This is easily explained by a glance at the geological map, for this catchment-area is hemmed in on the east by a hard ridge from 500 to 1000 feet in height.

The horse-shoe water-parting is intersected by passes or gaps at one or two points, and in the olden days these were strategic points or centres of weakness through which migrations and invasions successively took place. Later, they became the tracks of roads, and to-day the main traffic lines use them. The chief are the Tyne Gap, which is continued in the Irthing, itself a beheaded tributary stream of the Tyne now draining to the Solway; the Deadwater Gap between Larriston Fells and Peel Fell, used by the North Tyne and Liddel Water; and the Byrness Gap, used by the Redewater and containing Otterburn and Reidswire Raids. Lastly, there is the area of comparatively low contour, which has been the scene of many a bloody conflict (notably Flodden), lying between the Bowmont Water and the sea.

The chief rivers of the county are the Tyne, the Tweed-Till, the Coquet, the Aln, the Wansbeck, and the Blyth.

The Tyne rises in two main streams, the North and South Tyne, the former issuing from between Caplestone Fell and Hogswood Moor in a complex of burns, notably Kielder Burn, Scalp Burn, and Lewis

Burn. The South Tyne rises in Cumberland in the mountains east of Cross Fell. The North Tyne flows in a south-east direction for 32 miles and is joined along its course on the left bank by the Kielder Burn at Kielder, the Tarset Burn at Tarset, and the Redewater at Reedsmouth. It joins the South Tyne just below Warden,

Ward Hill and the North Tyne near Chollerford

north-west of Hexham. After entering the county from Cumberland the South Tyne flows due north past Alston in a pretty valley till it reaches Featherstone. Here it bends round towards the east and at Haltwhistle it has had a course of 19 miles. Below Bardon Mill it is joined on the right bank by the river Allen, a tributary due to the confluence of the East and West Allens, flowing from

Allenheads and Coalcleugh respectively. Gaining in width and depth as a dark, rolling stream, tumultuous when in flood, it passes Haydon Bridge and curves again northwards to Warden Mill, a distance from Haltwhistle of 14 miles. The united Tyne is joined on the right bank at Corbridge by the Devil's Water. Its valley now consists of one long series of fertile "haughs," or old flood-plains, and the river is only joined by one more

Meeting of the North and South Tynes, near Warden

important tributary, the Derwent at Derwenthaugh. From this point onwards to the sea, a distance of nearly 14 miles, it is one vast sequence of factories, chemical works, shipyards, smelting works, staithes, and docks—all summarised in the term "Tyneside."

The Tweed forms part of the northern boundary of the county, and receives the Till with its feeders, the Bowmont Water, the river Glen, the Harthope Burn, the College Burn, and the Breamish. Like all Northumbrian streams when they pass from the torrential area to debouch upon the plains, they form fertile flood-plains with many windings. The Breamish rises at Scotsman's Knowe in the Cheviot, near the source of the Harthope Burn which flows *via* Wooler to the Till. It passes

The Allen at Staward

Ingram, Hedgeley, Chillingham, and Chatton, and joins the Harthope below Wooler to form the Till.

The Till flows at an average elevation of between 100 and 200 feet for a distance of about six miles, and owing to its very low gradient is characterised by fine "oxbows" and windings. It passes Ford and Etal and joins the Tweed at Twizel Castle.

The Aln (16 miles) rises near Alnham (600 ft.) and flows *via* Whittingham, Hulne Park—near which it is joined by the Eglingham Burn—Alnwick, and Lesbury, and enters the sea at Alnmouth.

The Coquet, 40 miles in length, rises in Thirlmoor near Outer Golden Pot (1400 ft.), and is joined by six rapidly-flowing burns from the Windy Gate area. It flows south-east to Linsheeles, above which it is joined by the Usway Burn, direct from the Cheviot. It passes Alwinton, Harbottle, and Holystone in order, widening and losing in gradient all the way. About 2½ miles below Holystone it turns north-east to Rothbury, being joined on the left bank by the Wreigh Burn near Thropton. This is a region of wonderful windings and continues so for nearly 12 miles. From Rothbury the Coquet flows east-south-east for five miles and then turns north-east towards Felton, near which it is joined by the Swarland Burn. With a still zigzag course it passes Guizance and enters the sea at Amble.

The Wansbeck, 23 miles in length, rises in the Crag-lands just where a faint divide separates its head waters from the North Tyne basin. It is the overflow of Sweethope Loughs, which are 950 feet above the sea. It flows *via* Kirkwhelpington, Angerton, Mitford (where it is joined by the river Font), Morpeth, and Bothal, and enters the sea nearly two miles east-south-east of North Seaton.

The Blyth, which has a course of only 20 miles, rises in the Blackhill (739 ft.) north-west of Kirkheaton, and flows past Belsay, being joined by the Pont near Carter

Moor. The Pont has its source near Little Whittington (700 ft.) and passes Matfen, Stamfordham, and Ponteland. The united stream traverses the lovely Vale of Stannington and Plessey Woods, passes Bedlington, and enters the sea at Blyth.

Northumberland, with only a moderate rainfall and

Blyth Harbour

no distinctly rugged scenery, is devoid of large lakes. There are, however, some twenty water-areas in the county known as lakes, loughs, tarns, ponds, and reservoirs, many affording fine water-supplies for the industrial areas. The finest lakes are comprised in the group lying near Borcovicus, and include Greenlee Lough, Broomlee

Lough, and the beautiful Crag Lough. Grindon has been drained away.

Lying to the east of the North Tyne, in the limestone Crag-lands, is another group, including Sweethope Loughs, Colt Crag Reservoir, and East and West Hallington Reservoirs, which belong to the Newcastle and

**Crag Lough and the Whin Sill at Hot Bank**

Gateshead Water Company. This Company also owns the Catcleugh Reservoir lying at the head of the Redewater and at the foot of Catcleugh Shin (1742 ft.). It is the largest sheet of water in the county, being over $1\frac{1}{2}$ miles long by nearly half a mile in width. A group of five small tarns occur in the grounds near Rothbury. The Borough of Tynemouth obtains its supply of water from the river Font near Rothbury, where a huge reservoir has been constructed with a capacity of 720,000,000 gallons.

## 6. Geology and Soil.

Thoroughly to understand the significance of the physical geography of any district one must have some conception of its geology, for the surface-features, drainage-areas, climate, and industrial prosperity are intimately connected with the rocks of which the district is composed. Geology is the science of rocks, and it must be understood that soils and clays and other soft strata are to the geologist no less "rocks" than is the hardest granite. Broadly speaking, rocks can be classified into two main groups: (1) Igneous rocks, like granite, lava, and volcanic ash, and (2) Sedimentary rocks, which are caused by the disintegration of igneous rocks, and generally laid down in more or less horizontal layers under water. Sedimentary rocks very often contain fossils, which act as labels to them, indicating their relative age by their order of superposition. If the earth's crust had always been stable we should expect to find the oldest sedimentary rocks at the base and the younger ones above. But it has passed through many vicissitudes; whole countries have been elevated or depressed at different ages and consequently horizontality in sedimentary rocks is the exception and not the rule. Beds have been folded and tilted just as we may fold or tilt the pages of this book. Thus old and young rocks may lie at the surface together. Then their *outcrop* or *edge* or *strike* may be worn down by the action of rain, wind, or waves, and newer deposits formed upon them, as at Tynemouth Cliff. On folding, the beds may form an *arch* or *anticline* and thus *slope* or *dip* away in two directions. Very frequently the strata are snapped across or *faulted* so that the beds at one cheek of the fault do not at all correspond with those at the other. Now these faults form a very important feature in Northumbrian geology, as is shown, on the coast from Tynemouth to Blyth. At Cullercoats the greatest is seen in the 90-Fathom Dyke Fault, and on glancing at the map at the end of this book we can see how numerous they are. In fact, it may safely be said that much of the mining prosperity of Northumberland is due to these faults, for, had they not buried the coal seams deeper, the latter must assuredly have been planed off in the great Ice Age as they were in Ireland.

Igneous rocks have at one time been molten and have either been poured out at the surface as lavas, or injected between other beds and solidified there along or across bedding planes or in cracks or faults. Volcanic ash is the material thrown out from a crater and forms in greater or less degree the volcanic cone. These ashes may be in regular sloping layers and very often may resemble stratified rocks, particularly if they were deposited in water and contain fossils. Northumberland whilst built up in the main of sedimentary rocks, i.e. limestones, sandstones, shales, and coals, contains remarkable masses of igneous rock, notably the Cheviot and the

Great Whin Sill, than which, indeed, there is no better example in Britain. Cheviot is an old volcanic stump and the Whin Sill is a basalt or solidified lava injected between the layers of sedimentary rocks.

So great are the forces at work in the earth's crust that few of the beds are in their original state. Masses, in moving along the cheeks of faults, other areas being upheaved, cause strains or pressures to act upon the sediments, compacting, hardening, and fracturing them in a remarkable manner. Thus it is that sand becomes sandstone and mud is converted into shale. Very ancient deposits of shale or mud when compacted by very great lateral pressure split up into fine hard layers known as slate. Other cracks caused by pressure are seen in quarries and are called joints. These are very well shown in the Fell-top Limestone at Harlow Hill. It is along these joints that quarrymen drive wedges in loosening the rock.

If we could arrange all the rocks of England into horizontal beds they would form a mass probably from 50,000 to 100,000 feet thick. This is divided into four main groups or periods—Primary, Secondary, Tertiary, and Post-Tertiary or Quaternary. The Primary are the oldest and the Quaternary the youngest. The table on the next page gives this classification in all its details, and those beds in it which are Northumbrian are printed in heavy type. With this and the map before us we notice that, with the exception of the Recent Glacial Beds, they are entirely confined to the Primary. With this outline of the chief principles which are essential rightly to

| Systems | Primary Divisions | Subdivisions | Features |
|---|---|---|---|
| QUATERNARY | RECENT | Recent deposits | Alluvium, etc. |
| | PLEISTOCENE | Metal Age Beds | Valley and cave deposits |
| | | Neolithic Age Beds | |
| | | Palaeolithic Age Beds | |
| | | Glacial Age Beds | Striated boulders |
| TERTIARY | PLIOCENE | Cromer Beds | Chiefly sands |
| | | Norwich Crag | |
| | | Red ,, | |
| | | Coralline ,, | |
| | OLIGOCENE | Hempstead Beds | Sands, clays, limestones, deposited in seas, rivers and estuaries |
| | | Bembridge ,, | |
| | | Osborne , | |
| | | Headon ,, | |
| | EOCENE | Bagshot Sands | Sands and green clays |
| | | London Clay | Shallow-water clay |
| | | Woolwich and Reading Beds | Clay and pebbles |
| | | Thanet Sands | Sands |
| SECONDARY | CRETACEOUS | Chalk | Chalk |
| | | Upper Greensand | Sands |
| | | Gault Clay | Clays |
| | | Lower Greensand | River and estuarine deposits |
| | | Wealden Clay | |
| | JURASSIC | Purbeck Beds | Clays, sandstones, "egg-stone" limestones; highly fossiliferous |
| | | Portland ,, | |
| | | Kimmeridge Clay | |
| | | Corallian Beds | |
| | | Oxford Clay and Kellaways Rock | |
| | | Cornbrash | |
| | | Forest Marble | |
| | | Great Oolite | |
| | | Stonesfield Slate | |
| | | Fullers' earth | |
| | | Inferior Oolite | |
| | | Lias, Upper, Middle, Lower | |
| | | Rhætic | Highly fossiliferous |
| | TRIASSIC | Keuper Marls | Green and red or variegated marls and sandstones with gypsum and salt |
| | | Keuper Sandstones | |
| | | Upper Bunter Sandstone | |
| | | Bunter Pebble Beds | |
| | | Lower Bunter Sandstones | |
| PRIMARY | PERMIAN | Magnesian Limestones | At Tynemouth and Cullercoats |
| | | Marl Slate (*a*) | (*a*) contains fish |
| | | Yellow Sands (*b*) | (*b*) a desert sand |
| | CARBONIFEROUS | UPPER. | Marine, land, and fresh-water beds, sands, grits, shales, coals |
| | | Coal Measures proper | |
| | | Gannister Beds | |
| | | Millstone Grit | |
| | | LOWER. | Limestones, sands, shales, coals |
| | | Bernician Series | |
| | | Tuedian ,, | |
| | | Basement Beds | |
| | DEVONIAN OR OLD RED SANDSTONE | Upper | Red sands, shales, limestones, desert |
| | | Middle | |
| | | Lower ?? | |
| | SILURIAN | Ludlow Beds | Red sands, shales, limestones, flagstones |
| | | Wenlock Beds ?? | |
| | | Llandovery Beds | |
| | ORDOVICIAN | Bala or Caradoc Beds | Black shales, flags, thin limestones and sandstones |
| | | Llandeilo Beds | |
| | | Arenig Beds | |
| | CAMBRIAN | Tremadoc Beds | Slates, flags, shales |
| | | Lingula Flags | |
| | | Menevian Beds | |
| | | Harlech ,, | |
| | PRE-CAMBRIAN | Unclassified barren of fossils and only in patches | |

understand the geology of Northumberland we shall now give a brief epitome of it.

It is simple in its main essentials. The beds as a whole dip seawards, so that if a traveller were to walk from the Cheviot to the Tyne, he would successively meet with younger formations. He would thus pass over Igneous, Silurian, Carboniferous, and Permian strata, every now and then encountering glacial drift in all its

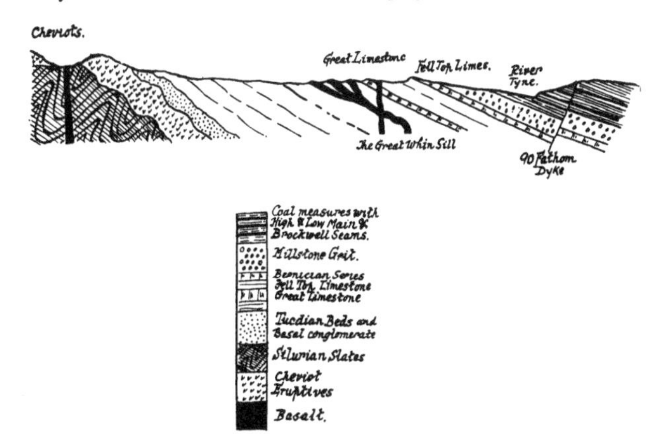


vagaries. We shall readily see how little of the series is represented in this county, and the neighbouring counties do not take us much further. On examining the map, it is at once apparent that the strike of the beds is arranged concentrically around the Cheviot as centre and the general dip is to the south-east and east.

The figure on this page shows what is known as a geological section drawn across the country from the Cheviots to the Tyne, and exhibits the arrangement of the rocks in the county as they would appear were a cutting made along this line. The oldest group is called the Lower Silurian. These are marine beds consisting of clay-slate and grits, and occur at the head of the Rede-water and to the north-east of it as tiny patches, sole relics of an age when the Cheviots first began to emit their lavas. The volcanic rocks of the Cheviots had already been deposited when Old Red Sandstone times had arrived. Following these are the basement con-glomerate of Upper Old Red Sandstone age, and the Lower Carboniferous beds deposited around the base and flanks of the old volcano and containing fragments of Silurian and Old Red Sandstone rocks. All through the period of the slow accumulation of the Carboniferous rocks which crept around the Cheviot mass, volcanic quiescence reigned, until towards the end of the period when it is probable that again there was a display of activity in the Cheviots. A huge sheet of lava or basalt was injected between and often through the Lower Car-boniferous rocks, and this lava doubtless crept along many of the fault planes and formed what are known as dykes. The main result of these forces was that the Car-boniferous beds were no longer horizontal but dipped towards the sea just as they do to-day. Towards the end of the Carboniferous period the country was slowly sinking, and the up-turned edges of these beds were planed down by the sea. On the shore of this sea rolled pebbles were formed, to be converted later into a con-glomerate. Sandstones, shales, and limestones of the Permian were deposited in this sea, which extended northwards as far as Blyth and west to about Killingworth. Now the third period of Cheviot activity arrives. The great faults like the 90-Fathom Dyke, which runs right across the county from Cullercoats, *via* Denton Burn, Greenside, and Whittonstall, were formed at this time. The Lower Carboniferous beds (Tuedian) stretch from Berwick, through Wooler, and skirt the Cheviot mass to the head of the Redewater. The Bernician Series containing the Fell-top and Great Limestones stretch from the Tweed to the Aln. From the Aln the Fell-top Limestone passes across the county parallel to the Great Limestone and crosses the Wansbeck west of Morpeth and the Tyne east of Corbridge. It then strikes due west, forming the high land to the south of the Tyne Gap. This series contains the great Whin Sill, curving from the Farnes to Bamburgh Castle past Alnwick, Rothbury, Chollerton, and the Northumbrian lakes to strike down the Pennines, reappearing again in the Tees at High Force. This is one of the outstanding features of Northumbrian geology, and its towering crags, capped by the Roman Wall, are most impressive. From Aln to Coquet the Millstone Grit or the Grindstone "Sill" is seen. It passes through Morpeth, crosses the Blyth and passes over the Tyne to the east of Corbridge. The rest of the county is occupied by the true coalfield capped by the Permian formation. The chief coal seams can be well seen in the cliffs, and others are worked in the adjacent collieries, i.e. the High Main, the Low Main and the Brockwell. The field is honeycombed with a network of east and west faults which cut it up in a remarkable way. The Permian beds occur in small but very interesting patches at Seaton Sluice, Cullercoats, and Tynemouth. The chief members are desert sand, the famous fossil-slate containing wonderful remains of fish, and the Magnesian Limestone, which forms the headland of Tynemouth and the remarkable cliffs and stacks of Durham.

The Permian Cliff, Tynemouth

Covering the main part of the county like a mantle is the most recent deposit, the Boulder Clay and Drift. It is in the main a homogeneous brownish or bluish clay laden with polished and striated boulders usually foreign to the district where they occur. These are relics of the Ice Age. It attains a thickness of 100 feet on the north bank of the Tyne at Tynemouth and forms large cliffs. The Whin Sill near Bamburgh, as well as the Rothbury Grits, show excellent striae in their surfaces under this clay. Boulder Clay contains very little fossil life. It is often overlain by drift sands and gravels. These are widely distributed in the county and largely fill the valleys of the Tyne, Wansbeck, and Aln. Still more recent deposits overlie these, notably the soils which vary in different areas and often partake of the nature of the bed rock. In the haughs (Ger. *haff*) on the river banks, alluvium occurs, forming fertile home-lands. Moor-peat cloaks even the highest hill-slopes and bog-peat fills the shallow depressions underlain by impervious beds.

## 7. Natural History.

Northumberland has been prolific in natural historians, and this is no doubt due in no small measure to the environment of the people. On all sides the fauna and flora are diversified in type and this has led to the establishment of societies and field clubs in the county. Naturally, then, we can but hope to skirt the fringe of this subject and leave the enquiring student to wander at liberty over the haunts of the brothers Hancock. Perhaps the best way to become introduced to the natural history of the county is to visit the Hancock Museum at Newcastle and there to contemplate the life work of these brothers. Plants, animals, birds, and fishes have all received special treatment. About four hundred is, according to the latest authorities, a rough estimate of the

Pinnacle Rocks, Farne Islands

(*The summits are crowded with guillemots and other sea birds*)

number of species of birds recorded in Britain, and quite two-thirds of these are frequenters of our county. Only Norfolk's avifauna rivals this, and its richness in birds has afforded ample scope for its study. This surprisingly great number of species of birds in Northumberland is no doubt due in the main to the diversity of physical feature giving rise to distinct botanical and ornithological provinces. The long seaboard lies in the direct line of flight of the migratory birds from northern latitudes. The coast in many parts is very high and rugged, and affords protection in the breeding season. Moorland, bog-land, and sand beaches all have their distinct avifauna. The higher elevations like the Cheviots and Simonside give rise to a sub-Alpine type, and here for a long time the eagle, peregrine falcon, and raven held sway (Raven's Crag, Simonside). The principal rivers like the Tyne and Coquet are favoured haunts of birds. Northumberland also lies within easy reach of the breeding stations of the sea-birds of the south of Scotland, and our number of annual visitants is increased by the proximity of the elevated moorlands and wild mountains of Cumberland and Westmorland. But it is the Farnes which surpass all other areas in richness of bird life. In this limited area of a few basaltic islands no less than fifteen or sixteen different species of sea-fowl breed, notably the eider-duck, puffin, razorbill, guillemot, cormorant, roseate tern, arctic tern, common tern, Sandwich tern, lesser black-backed gull, kittiwake, herring gull, ring dotterel, and oystercatcher. In the breeding season it is most difficult to pass over the nesting area without treading upon the eggs of some of these birds, and all the lofty pinnacles of the Whin Sill are covered with bird-lime. It is on these that the guillemots breed. Birds are classified into casual visitants, of which we get about 80; autumn or winter visitants about 54; residents about 92, and spring and autumn migrants about 40. Catalogues of birds show

**Eider Duck and Nest**

many rare species, but these need not be considered here.

About the animals proper little need be said. The badger, fox, and otter are still to be met with in the wild state. At Chillingham the famous White Cattle, figured on p. 72, supposed descendants of the aboriginal herds of Caledonia, are bred and protected.

With reference to the flora the distribution of species is controlled by many factors, such as the kind of rock and the ascending limits of temperature. This is well shown by noting the heights at which cornfields cease. Above this limit the flora usually assumes the nature of heath-land, characterised by heathers (*Ericae*), and bracken, bilberry or whortleberry (*Vaccinium Myrtillus*) and sphagnum moss and reed grass in the vicinity of pools and well-heads. This, then, is broadly the kind of vegetation covering the larger proportion of the upland parts of the county. It is also important to note the restricting influence of the rock upon the type of flora, though it is somewhat rare in this county, for out of some 800 species only about 50 show this restriction. It does happen, however, in limestone rocks, when we find dry-loving or xerophilous species living on limestones between claybands. Generally there are two main types of vegetation :—

1. That which lives best on rock which readily disintegrates to form soil, and is most common in Northumberland.

2. That which lives best on rock not readily disintegrated.

The first class of rock includes shales and sandstones, and the second limestones, basalts, and granites. Now in this respect the geological map is misleading, for whilst we note that a considerable proportion of the county is in a limestone series it nevertheless does not give rise to the characteristic scenery of limestone lands. As the rock does not possess in the main the essential limestone character we therefore find that the flora of Northumberland is mainly of the first type. The following list may be taken as typical of a Northumbrian hill of type 1 :—

| | |
|---|---|
| Mat grass | (*Nardus stricta*) |
| Hare's-tail cotton grass | (*Eriophorum vaginatum*) |
| Needle whin | (*Genista anglica*) |
| Bilberry or whortleberry | (*Vaccinium Myrtillus*) |
| Bog asphodel | (*Narthecium ossifragum*) |
| Grass of Parnassus | (*Parnassia palustris*) |
| Wood vetch | (*Vicia sylvatica*) |
| Fragrant orchis | (*Gymnadenia Conopsea*) |

On comparing Northumberland with other counties and with Europe in general we find that those species which are restricted to limestone bands occur with us in small number only; and those which are not so restricted preponderate, and owing to our humid and colder climate are much more developed. Dry-loving species occur in basalt and igneous regions. Thus the hairy rock-cress (*Arabis hirsuta*), sea campion (*Silene maritima*), knot-berry (*Rubus saxatilis*), liquorice vetch (*Astragalus glycyphyllos*), and kidney vetch (*Anthyllis Vulneraria*) may be taken as typical of the Cheviot. Though slaty rocks are rare in Northumberland it is quite certain that they absolutely exclude the limestone plants. Many species grow in sufficient quantity to produce a marked effect upon the scenery.

Near the ruins at Dunstanburgh the bloody cranes-bill (*Geranium sanguineum*) is common. Along the sea-shore the thrift or sea pink (*Armeria maritima*) is typical. Acres of blue-bells and garlic are to be found in the denes. The sand-dunes stretching from Seaton Sluice to Blyth are the habitat of various sword-grasses, with a variety of orchids, primulas, and violas. Like the beautiful denes of Mitford, Plessey, Swallowship, Holywell, Jesmond, and the bend of the Tyne at Chollerford, this spot is the rendezvous of numerous school parties and rambling clubs who visit them to study botany. The Chesters is famous for its Spanish erinus (*Erinus Hispanica*), which is a reputed relic of the *ala* of the Astures which garrisoned this camp.

Of marine mammals those found on our shore are best described as casual or accidental visitants. Porpoises and occasional rorqual and bottle-nose whales are recorded on the coast, and specimens are preserved in the Hancock Museum at Newcastle. The common hair seal is fairly abundant at the Farne Islands and they are frequently stranded at St Mary's Island and Tynemouth Cliff after violent storms. Several fine specimens captured under these conditions are preserved in the open-air rock tank at the Dove Marine Laboratory, Cullercoats. This fine institution with its modern equipment, destined to do so much for the local inshore fishermen, has by the patient research of those connected with it contributed much valuable information regarding the marine life of the north-east coast. Typical specimens are to be seen in the tanks. In his guide to the British Association Meeting at Newcastle in 1889, Professor Lebour had occasion to regret that "no good list exists of the fishes met with off the coast." Since then rapid progress has been made in recording and classifying the northern marine fauna.

## 8. Peregrination of the Coast.

The Tyne forms the southern boundary of the county for many miles. The mouth of this river has, through the agency of man, become practically a huge sea-dock, the finest harbour in northern England and one of the mightiest ports in the British Isles. Its entrance is protected by two magnificent piers, begun in 1854 and not completed until 1895. The North Pier, a great stone structure forming like its fellow an excellent sea promenade, was breached by a storm of unusual violence in 1897, and the reconstructed pier took 12 years to complete. It is nearly a mile long, and the South Pier, though considerably longer, is by no means as imposing. The Tyne entrance between these piers is about 400 yards wide, and is indicated to mariners by two lighthouses. The North Pier light is a fine granite structure, reached in stormy weather, when the seas break over pier and lighthouse, by means of an alley-way inside the pier. It shows a "grouped flashing white light" with three flashes every ten seconds. The South Pier has an occulting light eclipsed for two seconds every ten seconds. The entrance is indicated in foggy weather by a fog-horn on the North Pier with one blast every ten seconds, and a fog-bell rings on the South Pier every thirty seconds. Three other lighthouses, the Groyne on the Herd Sands visible for seven miles, and the High and Low Lights of North Shields, visible respectively 16 and 13 miles, show the importance of this harbour and the dangers which beset shipping making use of it in stormy weather.

Breach in Tynemouth Pier

The Municipal Borough of Tynemouth is so called from the pretty village of that name, beloved of Harriet Martineau, which is situated upon the fine promontory of Tynemouth Cliff. The busy industrial town of North Shields takes its name from the fishermen's "shiels" or huts which once occupied the site. Between the two lies the treacherous reef known as the Black Middens, the scene of many terrible shipping disasters. Rising above them, as he towered above his contemporaries, is the fine Lough monument of Lord Collingwood, with four guns of the "Royal Sovereign" on the base. The cliffs of Tynemouth, surmounted by a modern fortress, are over 100 feet high, and are of interest to the geologist as one of the few relics of the Permian age in Northumberland.

Stretching from the Tyne to the Blyth, Lower Coal Measures are met with at intervals in sections of great interest and considerable complexity but, as in the case of other parts of Northumberland, long stretches of dismal boulder-clay cliff intervene. Large expanses of sandy beach occur at Tynemouth, Cullercoats, Whitley, and Seaton Sluice, and afford delight to thousands of trippers in the summer months.

Cullercoats is a quaint fishing village, which was formerly important for the shipment of coal when the privileges of Newcastle debarred North Shields from exporting her produce from the Tyne. It is pleasantly situated on a fine little bay protected by two breakwaters. Here again the Permian formation affords an interesting section in the sandstone cliffs. Huge fractures in the earth's crust, known to the geologist as faults, form one of the chief features of the coast from this point northwards to Seaton Sluice. Admirable sections illustrative of the sequence of many of the Northumbrian coal seams are also met with.

On Brown's Point, north of Cullercoats Bay, stands one of the first and best types of wireless-telegraphy stations, which is in constant communication with the

Table Rocks, Whitley

Continent. A fine sea promenade extends on the top of the cliffs—locally called the "banks"—all the way from Tynemouth to Whitley Bay, affording interesting rock scenery and sea views.

Whitley Bay, a watering-place of rapid growth, much visited by trippers in the summer, lies to the north of Cullercoats and is well known for its Table Rocks, sea-bathing, and golf-links. A long stretch of fine sandy shore, backed up by a clay-cliff covered with blown sand, extends as far as St Mary's Island. This blown sand bears striking testimony to the prevalence of the north-east gales in the winter months. At intervals the cliffs have been beautified and laid out as gardens by the respective local authorities, and it has been found necessary

St Mary's Island

to protect the shrubberies from the drifting sand and biting wind by means of wooden screens. Here we may note how the glass panes of the lamps on the promenade are frequently etched by the sand-blast on the north-east or windward side.

St Mary's Island, once the seat of a Chapel of Ease of Tynemouth Priory, and the scene of many shipping disasters, is an interesting outlier of Carboniferous rock situated at the northern extremity of Whitley Sands. It possesses a fine lighthouse with a new flashing light, visible for 17 miles, established by the Trinity House to replace the old red light which was for so many years the prominent landmark at Tynemouth Castle. High cliffs stretch from here to Seaton Sluice, and the quaint village of Hartley with its red-tiled roofs is passed on the way. These cliffs, silent witnesses of terrible disasters, are succeeded by another stretch of sand and low cliff. It is interesting to note at this juncture how often the deeply-recessed bays coincide with the less resisting rock of the series.

Seaton Sluice is the most noteworthy instance on the north-east coast of decayed industry. Its once important harbour is now dammed up and bears sole testimony to the enterprise of the Delaval family, who did so much for the well-being of the place. The whole of this coast is characterised by a series of prettily-wooded streams and denes, which run down almost at right angles to it. The trees in Seaton Sluice Dene, easily visible from the coast, lean away from the north-east and indicate the direction of the dominant coastal winds. The wind-swept area between Seaton Sluice and Blyth is covered with accumulations of drifted sand or sand-dunes having a distinctive flora, and from the Tyne to the Blyth landslips form a remarkable coast feature. The major slips occur after periods of excessive rainfall and were common in October, 1903. Another interesting feature is the regularity with which certain sand-covered areas of coast are periodically cleared by the variation of the currents, which at certain times deposit huge masses of seaweed above mean tidal limits. This seaweed frequently decomposes, constituting a nuisance if the locality be near a town or village, but investigations have revealed it to be quite an important source of potash manure, and it will probably be extensively utilised both for this and in the chemical industry.

Blyth, nine miles from North Shields, is a safe and important harbour opening to the south-east and protected by two piers. The East Pier is a continuation of a spit of land projecting southwards from Cambois. This harbour has gone ahead with rapid strides. Its entrance is made easy to mariners by means of six lights and a fog-bell. Blyth owes its importance primarily to the large group of collieries situated in the immediate vicinity. Shipbuilding, coal-exporting, and fishing, however, are all responsible for the rapid advance it has made.

After passing the mouth of the Wansbeck, in the course of which are Morpeth and many collieries, we reach Newbiggin, which occupies a pleasant position four miles north of Blyth on a little bay between the Spital Carrs rocks to the south and the Outer Carrs to the north. It is a large fishing village and a popular watering-place, and was formerly of considerable importance as a corn-exporting port. The church of St Bartholomew forms a very conspicuous landmark for sailors, and occupies a prominent position on the rugged promontory of Newbiggin Point.

The coast now resumes its general north-north-west trend, and about a mile to the north we come to the Lyne sands and the small hamlet of Lynmouth. The rocky character of the coast continues past Snab Point to the fishing village of Cresswell, from which extends the fine semilunar Druridge Bay, five to six miles across, bringing us to the Bondicarrs rocks and the fishing village of Hauxley Haven with its rocky entrance. Just to the

**Baiting the Lines, Newbiggin**

north of this lies the Carboniferous outlier of Coquet Island, due east from the harbour of Amble.

Coquet Island, a mile off shore, has an area of nearly 16 acres and was formerly the site of a Benedictine monastery. The fine lighthouse has been erected on the site of the ruins of an old tower. The early establishment of this light is in itself indicative of the dangers of the coast. It is intermittent in type and is visible for 14 miles, and an explosive fog-signal every $7\frac{1}{2}$ minutes warns the mariner in thick weather. Amble lies at the mouth of the Coquet River, and combines a coal trade with its fishing, being near the group of collieries in the north part of the coal-field. Sea-birds make the island and the adjoining coast a favourite haunt. Two piers undergoing extensive alterations protect this rising harbour, and two important lights on the pier-head guide the traffic.

Alnmouth Bay, the duplicate of Druridge Bay, now lies to the north. Alnmouth, four and a half miles north of Amble, stands at the mouth of the Aln, and is a holiday centre with extensive golf-links and a fine foreshore. Cliff scenery again becomes one of the coast features, due to the change in the geological section. Though once important in the corn-exporting trade, the little watering-place to-day is merely a small fishing centre. Sand-dunes with their typical vegetation again appear. The river mouth is interesting as having been diverted by the breaching of a sand isthmus which in the last century joined the mound known as Church Hill with Alnmouth. Church Hill is of great antiquarian interest. Continuing northwards for about three miles we reach the old fishing village of Boulmer, notorious in the days of smuggling. This village, like Newbiggin and Cullercoats, is quite typical. Its old cottages, cobles, quaint people, and fine coast form a striking picture. Advancing northwards we pass the cape of Longhoughton Steel, Howick Haven, and Seahouses, and thus arrive at Cullernose Point before reaching the fishing village of

Craster. The geology now takes on an interesting character and the vagaries of the sandstones of the Howick coast are well worthy of note. But it is at Cullernose Point, just beyond Howick, that the scenery becomes grand, due to the fact that at this point the Great Whin Sill cuts the coast. Here the basalt with its rude columns rises to over 120 feet. Fault phenomena again form a feature of the section, and in the vicinity also the influence of the contact of igneous and sedimentary rocks can be observed in the masses caught up by the Whin Sill. Whin Sill phenomena, so pronounced a feature of the coast further north, are lost to the south of this, because it is here that the sill trends inland for the North Tyne.

Passing on, we reach the fishing village of Craster a mile north of Cullernose Point, and fine cliff scenery is maintained until we get to Castle Point, with Dunstanburgh Castle capping it, at the south extremity of Embleton Bay. Freeman asserts that the site of this castle is grander than that of any other in Northumberland. It certainly is imposing, situated as it is on the top of the basaltic Whin Sill, which is at this point about 50 feet thick and rests upon the metamorphosed beds of the Carboniferous Great Limestone series. Denudation along the columns of the basalt produces blowholes analogous to those so remarkably developed on the Durham shore north of Marsden. The Rumbling Churn or Rumble Churn at Dunstanburgh is a fine sight in stormy weather, when the waves are driven with tremendous force into the cavity, spouting the spray to a great height. The manner in which the sea readily removes the softer Carboniferous rocks, thus undermining the more resistant basalt, is specially well seen here. On the foreshore the Saddle Rock claims attention. The limestone underlying the Whin Sill has here been folded into a fine though small saddle or anticlinal fold.

Embleton Bay is barely two miles in extent, ending by

Saddle Rock at Dunstanburgh

Newton Haven in a rocky foreshore, which yields in turn to the fine bay of Beadnell, of about the same size, with a fringe of links and a small fishing harbour sheltering at the northern extremity. From the south side of Beadnell Bay to the Farne Islands, a distance of some eight miles, the Whin Sill is entirely absent. On the other hand the Farnes are in the main composed of it.

Beadnell, like Newton, is occupied to a considerable extent in the shell-fish industry. The coast from here right up to Berwick has a fairly straight north-west trend, and just here is very typical, showing a variety of limestones, sandstones, shales, and coal. Crossing the sandy bay past Annstead the Snook is reached, with North Sunderland and North Sunderland Seahouses in close proximity, the latter a small port devoted to fish-curing and lime-burning. Being served by the North Sunderland Railway it has become the centre for the Farnes, boats and guides being obtained here. Some three miles to the north-west lies the village of Bamburgh, the historical associations and the romantic position of which have rendered it famous. It lies just abreast of the Farnes, which are easily visible. Here the Great Whin Sill reappears, and attains a thickness of 80 feet, resting upon sandstones and shales. Naturally such a formidable position was utilised in the early days as a site for a fortress, and it was long considered impregnable. It presents a fine landmark for ships. The Farnes rise from the sea impressive in their majestic loneliness, standing like grim sentinels, and most forbidding in time of storm. They are located to the mariner by the famous Longstone lighthouse, reminiscent of Grace Darling's heroism. The light is white, revolving every half-minute and visible for 14 miles. A siren is used in foggy weather.

Passing a series of dunes of blown sand half a mile from Bamburgh on the way to Budle Bay are the Harkess Rocks, which to the geologist are synonymous with all that is perplexing in the study of intrusive igneous rocks and the influence which they have upon deposits in contact with them. Beds of shale and limestone have been caught up in the basalt in a remarkable way as the whinstone was injected through the series.

Passing Budle Point the large stretch known as Budle Bay lies below, and above the village rise the Budle Hills which appear to be associated with the Whin Sill, which is now striking inland.

From Budle looking north the lonely level of the Ross Back Sands leads the eye up to Holy Island, with the Ross Links and the sandy projection called the Old Law bounding them to the west. From here the vast stretch of Fenham Flats and Holy Island Sands, five miles in length by two in breadth, intersected with meandering streams, lies before us, and it is at once apparent that Holy Island is only an island at high tide. The village of Beal, due west from Holy Island, is the best centre for those desirous of visiting the latter, the cradle of Christianity in Britain. The island can be reached on foot or by cart, the track being marked by a line of beacons and refuge boxes, which can be used if the traveller is overtaken by the tide.

From Holy Island northwards to Cheswick Black Rocks the five-fathom line makes a wide detour so as to embrace Holy Island. Consequently at low tide huge expanses of sand are laid bare, notably the Goswick and Cheswick Sands, and one cannot help thinking that here lies a wide field for the work of reclamation. Naturally the contour of the adjacent mainland is very low. To the north-north-west of the island lie several shoals with less than five fathoms of water, notably Spittal Hirst, Inner Hirst, North Tours, Tours, Parkdyke, and Wingate Reef. A similar series extends from Holy Island towards the Farnes, including Plough Beacon, Nicholas Rock, Goldstone, Guzzard, and Tree o' the

The Track to Holy Island

House, all lying across the entrance to Holy Island harbour.

At Goswick Station the railway line comes close to the coast, and five miles north of this Spittal and the mouth of the Tweed are reached.

The ancient and historic town of Berwick is on the north bank of the river, and the township of Tweedmouth with Spittal on the south. Just below the town ramparts of Berwick and in close proximity to the military camping ground of Magdalen Fields lie some fine natural sea baths, the like of which are only met with at the Table Rocks, Whitley Bay, near the Tyne. Berwick Harbour lies between Sandstell Point and the North Pier, which is about 900 yards long and was built 1810–1821. It forms not only an efficient breakwater but an excellent promenade. A red and white beacon indicates the harbour entrance, and has been placed on Callot Shad just abreast of Spittal Chemical Works. On the pierhead there are two fixed lights on the same tower; the high light, which is white, is visible for twelve miles, the lower or red light for eight miles.

Northumberland extends north of the Tweed to about a mile north-west of Marshall Meadows Bay, a coast-line of nearly four miles beyond the river. This forms the eastern boundary of the small County of the Borough and Town of Berwick-upon-Tweed. From the point mentioned the boundary strikes south-west to Mordington Church and then due south till it meets the Tweed. The land rises steeply from the coast and culminates in Halidon Hill (537 ft.) about the centre of this small detached county, which was constituted a separate county in the reign of William IV. "Berwick Bounds," or the Liberties of the Borough of Berwick, include an area of approximately eight square miles and extend northwards as far as Lamberton Toll. Paxton Old Toll lies on the western margin of this miniature county. About two-thirds of the land, or 3077 acres, belong to the Freemen.

Coastal losses are not so apparent in Northumberland as in other parts of England, but are distinctly noticeable at Percy Square, North Shields, where houses are rendered uninhabitable by the frequency of landslides, due to the peculiarity of the boulder-clay cliff. Sea walls to protect the promenades have been constructed at Tynemouth, Cullercoats, and Whitley Bay. The most notable case of reclamation is at the Fish Quay, North Shields, where the foreshore west of the Black Middens is being converted into a Fish Quay extension to cope with an increasing fish industry.

The tidal wave enters the North Sea from the North Atlantic between the west coast of Norway and the east coast of Britain. Its average rate in the offing rarely exceeds one and a half knots, but where constricted by a narrow channel—e.g. the Pentland Firth—it acquires a greater velocity, even attaining the rate of eight knots an hour on spring tides. It is the western branch of the North Sea wave, confined by the Dogger and other banks, which gives us the high water in our rivers and bays successively as it passes southwards towards the Thames. About three miles off Berwick it runs till 4 hours and 10 minutes after the time of high water at Leith and so on in sequence we find that at five miles off North Sunderland Point and at a similar distance to the south-east of the Staples the tide continues in flood 3 hours 25 minutes after high water at Leith. Off Blyth and the Tyne and at distances of five and four miles respectively the tidal wave runs to the southward till 3 hours 40 minutes after the time of high water at Leith.

Tidal effects are best observed in the Tyne. A survey of the Tyne made in 1813 recorded a depth of water not exceeding six feet on the Tyne Bar at low water. It also showed that the channel of the fairway had a minimum depth of four feet. The soundings of this river, it is hardly necessary to say, are very different to-day. For a distance of approximately 14 miles, by a slow process of dredging, widening, and removing constricting promontories, the Tyne harbour has been so vastly improved that, from the Tyne Bar for about three miles, there is a depth of water of 30 feet (mean) at low water ordinary spring tides. To attain this end it has been computed that roughly some 130 million tons of silt and other debris have had to be removed at a cost of £2,469,147. The labour involved in the removal of this alone is immense, to say nothing of the fleet of dredgers, hoppers both steam and wooden, and tugboats employed in conveying it beyond the three-mile limit. Whilst in the main this has had a beneficial effect upon the general prosperity of Tyne communities it has brought certain hardships upon places like North Shields and Tynemouth by the concomitant increase in the range of the tide caused by the removal of so much matter. Our illustration on page 74 shows the T.S.S. *Mauretania*, then the largest liner afloat, leaving the Tyne on October 22nd, 1907, and drawing about 33 feet of water. The improvements of the Blyth are similarly causing corresponding increments in the total tonnage and trade.

## 9. Climate.

Climate, by which we mean the average conditions of the weather of a region, is the resultant of many factors, the chief of which are rainfall and temperature. But so intimately linked are these factors that any consideration of one involves a study of the other. Thus it is that a study of temperature demands also a consideration of the position of a place, both as regards its latitude, altitude, and proximity to the open sea. Similarly, rainfall has associated with it the factors of barometric pressure, prevalent wind direction, and the disposition of the condensing surfaces of mountain barriers. Of places in the same latitude those which are far inland are more extreme in climate than those on the seaboard, or in other words inland places have a *continental climate*, while those near the sea experience an *insular climate*. The sea acts as a controlling factor, tempering alike both summer heat and winter cold. Thus we should naturally expect that places away from this influence would show a greater variation in their climate.

Lines joining up places which have the same temperature for any specified time are called by meteorologists isothermal lines. Such lines are drawn on maps for January and July, the coldest and hottest months of the year, and such a set of lines are shown on the annexed map. These are, unfortunately, rarely complete for any particular county, but what are shown are the results of many observations. It will be noticed that the two sets intersect in several points and at these

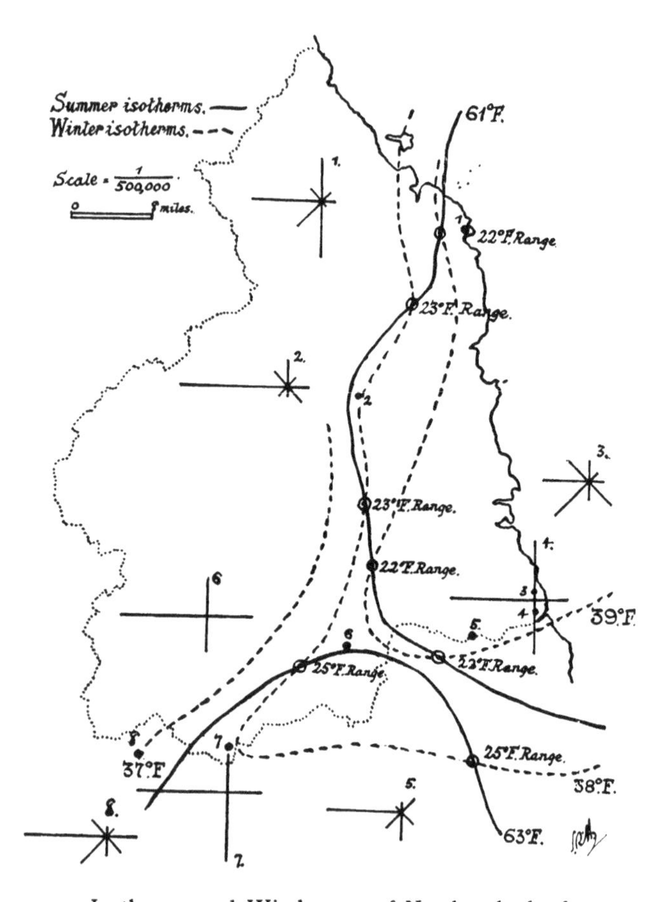


Isotherms and Wind-roses of Northumberland

(*To keep the map clear the wind-roses are placed slightly out of position. The centre of each rose should be over the point correspondingly numbered*)

points the difference between the summer and winter temperatures is given. This is called the *range*. A high range is characteristic of a continental climate and a low range of an insular climate. Thus the range of temperature for Cornwall is 20° F., for Northumberland 25° F., and for Vienna 39° F. The warming influence of the sea in winter is well shown in our map, for the temperature sinks as we pass inland from the sea. The influence of latitude is best brought out by the summer isotherms. Thus southerly places in summer are warmer than northerly, and in this respect Northumberland is like England as a whole. Our map illustrates clearly that the climate of the north-east coast is not nearly so severe as is commonly supposed. Places actually on the seaboard are in winter warmer than those inland. In this respect our coast is like the western and south-western shores of England.

Turning now to the rainfall map we find that it harmonises with what we know of contour and as we shall see presently with wind direction. Most of our weather comes to us from the Atlantic. The air, heavily laden with moisture from its passage over the ocean, meets with high land directly it reaches our shores, the Cornish moors, the mountains of Wales, or the fells of Cumberland and Westmorland, and at once parts with much of this moisture as rain. This is well seen on the annexed map, where it will be noticed that the heaviest fall is in the west and that it decreases with remarkable regularity until the least fall is reached on our eastern shores. The highest rainfall is thus in the highest land,

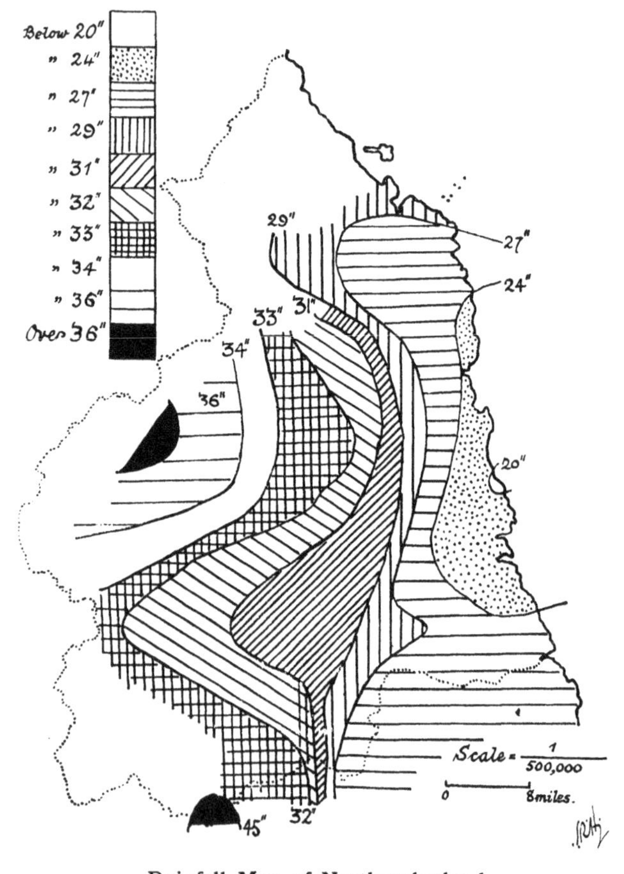


Rainfall Map of Northumberland

(*Showing the regular decrease in the fall from West to East*)

and here also rise the chief rivers. This bears interesting comparison with the "wind-roses" inserted for convenience upon the isothermal map. The lines in the wind-roses stand for actual directions and represent proportionally by their lengths the average number of windy days from each quarter. The greatest rainfall is in the Cheviot and at Allenheads, and the rainfall gradually diminishes towards Cresswell as a centre. The rainfall in this vicinity is amongst the lowest in Britain. Comparing with Wales, Killarney, and the Western Highlands of Scotland we find that whilst the rainiest belt in our county is some 40 inches per annum, in the districts named it reaches to between 80 and 100 inches. There also we find the maximum rainfall in the highest land.

Northumberland lies on the dry or "rain-shadow" side of the Pennines. Most British rain, as we have seen, is carried from the North Atlantic Drift by the west and south-west winds, and is distributed by the passage of cyclone after cyclone. From a consideration of the wind-roses it will be apparent that these winds bring our rain, deposit it upon the cold uplands, and pass over eastern Northumberland leaving it comparatively dry. But "high and low rainfall" does not necessarily mean that places have many or few rainy days. It frequently happens that in many places of very high rainfall it falls in a very short time. A rainfall map may therefore be misleading if one does not know the corresponding number of rainy days. Thus Newcastle for a certain number of years had an average number of 154 rainy days for an annual rainfall of 24 inches, Howick for a similar time had 150 rainy days for 22·9 inches, and Allenheads had 268 for 48·4 inches.

Climatology also involves a knowledge of cloud conditions and humidity apart from rainfall. Northumberland is not very well favoured in this respect, being singularly damp and cloudy, and the following statistics for August 1907 are given for comparison :—

| County | Humidity (per cent.) | Parts of Sky clouded out of 10 | Rainfall (in inches) |
|---|---|---|---|
| Northumberland | 81 | 8 | 2·5 |
| Lancashire | 79 | 8·5 | 5·4 |
| Dartmoor | 79 | 8·5 | 2·3 |
| Kent | 78 | 6·7 | 1·5 |

With reference to sunshine, records show that this varies very much all over the country, the greatest number of hours being registered on the south and east coasts of England and the least in the Southern Pennines near Manchester. Clacton-on-Sea in Essex registered 1945 hours of sunshire in 1906. Taking an average year, some parts of South England registered 1700 hours or more, while for the same year the record for Newcastle was 1000 hours and for Manchester 900 hours.

Severe frosts and heavy snowfall are not very common. Such weather does occur in the winter months, generally when Britain is under a pronounced barometric depression and a succession of cyclonic centres passes due south over the county, accompanied by regular blizzards and biting east and north-east winds. Severe north-east gales are frequent about September and October, causing distress to shipping on the open coast and much damage and loss all along the foreshore.

## 10. People—Race, Language, Settlements, Population.

We have no written record of the history of our land carrying us beyond the Roman invasion in B.C. 55, but we know that Man inhabited it for ages before this date. The art of writing being then unknown, the people of those days could leave us no account of their lives and occupations, and hence we term these times the Prehistoric period. But other things besides books can tell a story, and there has survived from their time a vast quantity of objects (which are daily being revealed by the plough of the farmer or the spade of the antiquary), such as the weapons and domestic implements they used, the huts and tombs and monuments they built, and the bones of the animals they lived on, which enable us to get a fairly accurate idea of the life of those days.

So infinitely remote are the times in which the earliest forerunners of our race flourished, that scientists have not ventured to date either their advent or how long each division in which they have arranged them lasted. It must therefore be understood that these divisions or Ages—of which we are now going to speak—have been adopted for convenience sake rather than with any aim at accuracy.

The periods have been named from the material of which the weapons and implements were at that time fashioned—the Palaeolithic or Old Stone Age; the Neolithic or Later Stone Age; the Bronze Age; and the Iron Age. But just as we find stone axes in use at the present day among savage tribes in remote islands, so it must be remembered the weapons of one material were often in use in the next Age, or possibly even in a later one; that the Ages, in short, overlapped.

Let us now examine these periods more closely. First, the Palaeolithic or Old Stone Age. Man was now in his most primitive condition. He probably did not till the land or cultivate any kind of plant or keep any domestic animals. He lived on wild plants and roots and such wild animals as he could kill, the reindeer being then abundant in this country. He was largely a cave-dweller and probably used skins exclusively for clothing. He erected no monuments to his dead and built no huts. He could, however, shape flint implements with very great dexterity, though he had as yet not learnt either to grind or polish them. There is still some difference of opinion among authorities, but most agree that, though this may not have been the case in other countries, there was in our own land a vast gap of time between the people of this and the succeeding period. Palaeolithic man, who inhabited either scantily or not at all the parts north of England and made his chief home in the more southern districts, disappeared altogether from the country, which was later re-peopled by Neolithic man.

Neolithic man was in every way in a much more advanced state of civilisation than his precursor. He tilled the land, bred stock, wore garments, built huts, made rude pottery, and erected remarkable monuments. He had, nevertheless, not yet discovered the use of the metals, and his implements and weapons were still made of stone or bone, though the former were often beautifully shaped and polished.

Between the Later Stone Age and the Bronze Age there was no gap, the one merging imperceptibly into the other. The discovery of the method of smelting the ores of copper and tin, and of mixing them, was doubtless a slow affair, and the bronze weapons must have been ages in supplanting those of stone, for lack of intercommunication at that time presented enormous difficulties to the spread of knowledge. Bronze Age man, in addition to fashioning beautiful weapons and implements, made good pottery, and buried his dead in circular barrows.

In due course of time man learnt how to smelt the ores of iron, and the Age of Bronze passed slowly into the Iron Age, which brings us into the period of written history, for the Romans found the inhabitants of Britain using implements of iron.

We may now pause for a moment to consider who these people were who inhabited our land in these far-off ages. Of Palaeolithic man we can say nothing. His successors, the people of the Later Stone Age, are believed to have been largely of Iberian stock ; people, that is, from south-western Europe, who brought with them their knowledge of such primitive arts and crafts as were then discovered. How long they remained in undisturbed possession of our land we do not know, but they were later conquered or driven westward by a very different race of Celtic origin—the Goidels or Gaels, a tall, light-haired people, workers in bronze, whose descendants and language are to be found to-day in many parts of Scotland, Ireland, and the Isle of Man. Another Celtic people poured into the country about the fourth century B.C.—the Brythons or Britons, who in turn dispossessed the Gael, at all events so far as England and Wales are concerned. The Brythons were the first users of iron in our country.

The Romans, who first reached our shores in B.C. 55, held the land till about A.D. 410; but in spite of the length of their domination they do not seem to have left much mark on the people. After their departure, treading close on their heels, came the Saxons, Jutes, and Angles. But with these and with the incursions of the Danes and Irish we have left the uncertain region of the Prehistoric Age for the surer ground of History.

It is certain, from the knowledge we have of the Roman Wall, that, after the granting of the marriage charter (a copy of which is preserved at the Chesters), the Roman soldiery intermarried with the British people, but owing to the state of servility of the latter people under Roman domination much of their strength of character was lost, and they thus fell an easy prey later to the incursions of the various Teuton hordes. Angles, Saxons, Danes, and Norsemen have all left their impress upon the physical type of the people, their customs, and their place-names, though naturally upon some more than others, according to their degree of establishment in our county. They in turn yielded place to the Normans, and after this there was a gradual fusion of types, though even now there are many primitive folk of quaint speech and customs, notably in the fishing communities and in places remote from towns. Another point worthy of note is the influence which intermarriage with Low Germans is to-day working in the seaport towns. This is readily recognised in the children in the seaport town schools.

The clue to the history of a settlement lies frequently in the place-names. A glance at a map of Mexico or the St Lawrence basin will show this. In this respect Northumberland is unique. As the westward pressure took place, the indigenous peoples and their successors were in turn vanquished and driven to the mountain fastnesses for security, a fact which may account for the more frequent occurrence of the so-called "camps" and stone circles in the hills. It is here also that we expect to meet with the oldest place-names, whilst the newer ones are found in the valleys or on the seaboard.

An analysis of the common place-names shown on an average map of Northumberland yields the following results. Out of 130, 82 are undoubtedly of Saxon origin, 30 are common to Anglian and Saxon, only three being essentially Anglian. If we include the doubtful *wick* and *ness* as Danish then ten are Danish, four are German, three Latin, and one Norman French.

The Northumbrian is descended in the main from the Angli (Bede). Danish settlements can be easily recognised by the yielding of the Anglian term *burn* to the Danish *beck* and by the place-name suffix *by*. Northumberland is, however, singularly free from Danish influence, as will be revealed by a close scrutiny of the language and place-names, and this agrees with history. The ultimate result of the influences of these varied peoples has been the production of a very remarkable dialect commonly known as "Tyneside" and characterised by the Northumbrian "burr." Border feuds, the frequent passage of troops, and the migration of Scots have not left the language unaffected, though it must be remembered that the Lowlander of Scotland is ethnologically a Northumbrian in type. The "burr" is known as the tonsil or uvular "r," and whilst not confined entirely to Northumberland is more widespread here than anywhere else. An investigation of the dialect has shown that if a line be drawn on the map passing through Berwick, Wark-on-Tweed, Wooler, Alwinton, Byrness, Falstone, Allendale, Blanchland, Ebchester, Birtle , and North Shields, it includes the area of the tonsil "r." It was commonly supposed that it came from the Danes. How then is it absent in the true Danish areas? It is now generally conceded that as a language-characteristic its date is much later. It was singularly rife in the time of Harry Hotspur.

We come across traces of pre-Roman settlements (e.g. in the name Barcombe—Borcovicus). The successive invaders sometimes occupied them, but no doubt founded many new ones, at first in the fertile valleys and later in the wilder areas. With the Normans came the settlement of homesteads and the parcelling out of estates.

The discovery of coal, its utilisation in the arts, and the inventive genius of many inhabitants, led in time to the centralisation of a huge population. The last census revealed the high density of the population of this county —345 to the square mile. It stands sixteenth in the country and this is apt to be very misleading, for if a density-of-population map be placed side by side with one of industry and coalfield it is readily seen that the concentration of population is almost entirely over the coalfields. Otherwise the strictly agricultural county is sparsely populated.

## 11. Agriculture.

The fundamental factors which determine to what extent a county shall be agricultural are its latitude, its altitude, its nearness to the sea, the character of its soil, and to a certain extent the natural bent of its inhabitants. The first three are rightly regarded as climatic factors and the soil has already been treated in the chapter on geology. The natural bent of a people is also important, for in the study of geography nothing is more apparent than the effect of national indolence or virility in determining position in the scale of peoples. Some win harvests from practically bare rock and others neglect the opulence of nature.

The present condition of the agriculture of Northumberland can be best understood from the diagrams at the end of this book together with the contrast afforded by the following table.

| | 1803 | 1912 |
|---|---|---|
| Sheep | 351,547 | 1,107,226 |
| Wheat | 39,245 acres | 5,179 acres |
| Barley | 21,881 " | 31,684 " |
| Oats | 71,803 " | 40,655 " |
| Horses | 27,069 | 18,488 |
| Pigs | 27,987 | 13,453 |
| Cattle | 76,640 | 27,919 |

To-day permanent grass covers 39 °/₀ of the county. Mountain heath and grazing land take up another 36 °/₀; 0·3 °/₀ is bare, fallow, or uncropped, and there is 3 °/₀ of woodland.

The land devoted to wheat is now less than one-seventh what it was in 1803 and it has passed into pasture with an enormous increase in the production of sheep. The quantity of oats has diminished more than half and barley has increased about half. A corresponding fall has taken place in horses, cattle, and pigs. This is in part accounted for by the growth of towns, due to industrialisation, and also to the vast development of wheat in Canada and cattle-ranching in the Americas.

On comparing Northumberland with other counties in the country in respect of sheep, it is at once seen that she stands easily first and is only approached by Kent and Lincoln. The cause is partly climatic and partly due to the high percentage of grazing and hay-land. The rainfall of Northumberland is fairly low and this favours sheep-rearing.

There are three large agricultural tracts, (1) the parallelogram from Tynemouth to Warkworth, Rothbury, and

Bywell, (2) the coastal plain from Rothbury to Alnwick, Belford, Lowick, and Twizel Castle, and (3) the basin of the river Till. The Lower Cheviots afford good grass land, but in the highest moors the land consists of desolate moss-hags, bogs, and scars which are practically worthless. The lower uplands are usually covered with a coarse type of wiry grass or "bent-hay."

Naturally the best crops are produced where the rigours of the north-easterly weather are not so keenly felt, i.e. in the sheltered valleys and along the fertile haughs. Boulder clay forms the greater portion of the soil, and it varies according to the nature of the bed-rock. Northumberland is not sufficiently dry and sunny ever to become a great wheat-producing county. The effect of altitude upon temperature and rainfall causes crops to be confined to valley bottoms, and thus the fells and heath lands are mainly pasture. Only by great patience, skill, and diligence can many parts of this county be made productive, and this it is without doubt that has contributed in no small degree to make the Northumbrian what he is to-day.

The demand for food in the industrial area has given an impetus to the rearing of cattle and sheep. There are many large sheep-markets in the county, particularly those of Rothbury, Hexham, Morpeth, Alnwick, and Wooler. The chief breeds are pure Cheviots in the uplands, black-faced sheep on exposed heath-clad heights, and crosses between Leicesters or Shropshires and Cheviots in the lowlands. Dairying and experiments in farming are conducted by Armstrong College and at the

County Agricultural Experiment Station, Cockle Park, Morpeth.

Farms are held under long leases. This allows the farmer to remain unhampered by the restrictions which result from short tenure. In some of the more elevated tracts the farms are more or less "strip-like" so that each possesses fell-top ground for sheep, slope-ground, and valley-ground for tillage. This is not so well marked as in Cumberland. Most farmers desire a good proportion of their land for pasture, in many cases as much as one half. Consequently in the valley farms, which are essentially arable, this is only rendered possible by the observance of some specific scheme of rotation of crops. The Berwick or five-course rotation is worked in some areas, though this varies considerably. In it corn is followed in yearly sequence with roots, corn, grass, and grass. However, where the land is poor in quality it is sometimes kept as much as three years under pasture before it is ploughed or not infrequently is allowed to lie fallow. The following rotation is carried out in some turnip lands and scarcely differs from the Berwick rotation. Oats is followed by turnips or potatoes, spring wheat or barley, clover or other grasses, and pasture.

Simplicity of character, hospitality, and self-reliance are features of the typical fell farmer. Constant contact with the inclement weather has made the shepherd a keen observer of nature and atmospheric phenomena, for he knows that blizzard or flood overtaking him suddenly may do him irreparable damage.

Coming to the farm-hands proper the nomadic habit

is quite distinctive and is evidenced by the periodical hirings held in all the chief market towns. The "village-feast," "merry-meeting" and "mell" or "kern" supper, rare except in the interior of the county, are relics of her ancient hospitality and rustic character. The spirit of co-operation between farmer and hind has resulted in the custom of employing men at "stock-wages." Under this system shepherds are allowed to keep at their masters' expense a dozen or so of ewes and a cow. They also have so many yards of potato land, a house, and a small money wage. This spirit is further exemplified in the custom of holding joint "clipping" (sheep shearing) and harvesting days, when by mutual arrangement an interchange of labourers takes place. A noteworthy feature of Northumbrian agriculture, and particularly near the towns, is the large number of women workers. Whilst the men are employed in shepherding, sheep-washing, clipping, ploughing, sowing, or reaping, the women frequently manure the fields, sow and pick potatoes, hoe or gather turnips, and cut food for folded sheep or cattle.

The inhabitants of the high valleys are direct descendants of the borderers or dalesmen, who were noted thieves and cattle-stealers. The result of this was the development of deadly border feuds. The rival chiefs on both sides of the border countenanced this predatory warfare and in the early days legislation was introduced holding these "dale" chiefs responsible for all the misdeeds of their kinsmen. There can be no doubt that pillaging conduced in no small measure to the erection of a large number of the "pele" towers for which our county is famous. Here cattle could be driven in time of warfare or feud. Cattle-breeding for food and dairying is fairly considerable, but the products are mostly consumed in the county. The chief cattle bred are Shorthorns and Galloways. The famous Chillingham herd of wild cattle, preserved at Chillingham Castle, deserves special mention.

Chillingham Wild Cattle

It has been rendered famous by the work of Bewick, the engraver. It is supposed that the animals are descendants of the aboriginal Caledonian herds. They are described in Bewick's *History of Quadrupeds*, and Bewick's "Chillingham Bull" of 1788 is a masterpiece of engraving. The Earl of Tankerville gave details concerning the herd before the British Association in 1838.

## 12. The Origin and Growth of the Tyne Ports.

By nature the Tyne is singularly blest, for close at hand and extending for miles in every direction are the materials requisite for building up a great industrial trade. For centuries they lay practically undeveloped, owing to the nature of the river and its bad and dangerous harbour. Its geographical position opposite the Baltic ports, its huge coal resources, the mechanical bent and energy of its people determined to a considerable extent that the Tyne should take the lead in Britain's industrial greatness. Truly has it been said that Tyneside was the cradle of nineteenth century progress. It was here that steam was first satisfactorily harnessed and applied for man's great gain. Here, too, originated many other inventions which have proved great factors in the civilisation of the world.

Let us look at the condition of this river as late as the opening of the Victorian era. At this period the Tyne was a placidly-flowing, tortuous stream, fordable at Newcastle, with scenery only now to be met with in the upper reaches. Below Newcastle plantations with country houses peeping out, green fields, and pleasant denes met the eye on both banks. This shallow and crooked stream was destined to be transformed, however, in the short space of a man's life, into a great artery of commerce. As late as 1849 there was practically no harbour; the average depth of the bar was but six feet, and steamers of 400 to 500 tons burden, after loading part cargo at Newcastle, had to obtain the remainder at North Shields in order to avoid the shoals and shallows. This same river was destined to send forth the *Mauretania*! At that time the Tyne harbour was notorious on the east coast for its danger and the Black Middens were only too well known. No docks and but little quay accommodation existed, and most of the cargoes had to be shipped by means of barges

The "Mauretania" leaving the Tyne

or "keels," the building of which was quite a thriving industry. These coal barges were exceedingly numerous, and so important a factor were they in shipping that cargoes were measured in "keel-loads." That the Tyne coal industry was developed at this time is well borne out by statistics, and the fact that collier brigs from Shields were to be met with in all parts of the world in the time of Collingwood is ample confirmation.

About 1849 there arose a feeling that this fine coal-field and centre of trade was seriously handicapped by the lack of a good harbour and river. The authorities of the city of Newcastle had not developed the latter. The result was that in 1850 the River Tyne Commission for the improvement of the river was formed. What a transformation has been wrought is to be seen to-day in this man-made river, with its magnificent stone piers, its fine depth of water, and its absence of shoals. Even the Black Middens have lost their terror. The piers themselves are a remarkable feat in engineering skill. Stretching respectively for a distance of 983 and 1717 yards into the sea, and costing £1,616,000, they form a fitting finish to the work of those who strove to make the Tyne a worthy channel for her mighty industry. They were commenced in 1854, when the foundation stone was laid by Mr Joseph Cowen, and they were not completed till 1895. This progress of the river and the attendant increase in the size and number of its factories and their output was a slow process, an unconscious growth. People grew up along with the changes and thought little of them till they chanced upon some old person who spoke of "the other days and the other Tyne," days of clear skies and green fields, not as now of smoke and grime. It seems incredible that such changes could happen in a few decades, though before the coming of the locomotive the Tyne had a large trade and took high rank as a British port. To measure the progress of the Tyne one but needs to record her yearly output of coal.

We have seen how the Tyne has grown; let us now

pass in brief review her engineering enterprises. Locomotives, as we shall see, originated and for many years were built here, and the tradition is still carried on in the construction of the best marine engines. Britain's commercial supremacy lies in her carrying trade, which is due in no small measure to the scores of tramp-steamers and colliers coming and going from our ports. Walker-on-Tyne, in 1842, saw the building of the first iron hull; and in 1852, from Palmer's works at Jarrow, the first iron-built, screw-propelled tramp, the *John Bowes*, was sent out. This period was perhaps the richest and most far reaching in the history of British invention. Armstrong, by his perfected application of the principles of hydraulics, by the designing and fashioning of his coil-welded, rifled guns, made the Tyne celebrated, and established a factory at Elswick which to-day numbers its workers by tens of thousands and has a river frontage of a mile. His revolutionary discoveries in ordnance changed the entire complexion of warfare. From Jarrow was launched in 1856 H.M.S. *Terror*, the pioneer in heavy, armour-plated battleships. It is not to be wondered at, therefore, that in the naval world the Tyne occupies a pre-eminent position. Moreover the steam-tug and the "oil-tank" or bulk oil-carrying steamer, both of which are important factors to-day in shipping, were essentially Tyneside conceptions, and the Tyne still retains its position as a builder of these vessels.

But even before the period of great invention the iron industry was established. There were then no iron ship-yards and engineering was in its infancy, yet cables, anchors, bridges, and iron plates were distributed all over the world from this river. Almost simultaneously with the rise in engineering, this trade declined and for many years there were dismal stretches of dismantled iron-works. Modern demands, however, are reclaiming these, and out of their ashes at Walker is rising what will be one of the largest yards on Tyneside. In 1850 the pottery, bottle and plate-glass, and chemical industries were at the zenith of their prosperity. The chemical industry alone remains the survivor of competition. Smelting of copper, lead, and iron was established and is still extensively carried on. All this enterprise and industry would have been of little avail but for the marvellous improvements in the river. Its course had by this time been straightened and deepened so that vessels of large draught could enter practically at any tide, and could discharge at the wharves at Newcastle or load at the staithes either at Dunston, Pelaw Main, Howdon Dock, or Tyne Dock. To-day the harbour is wide enough to hold a naval squadron and deep enough for Dreadnoughts and Cunarders—in fact the Tyne is only one huge sea dock.

Of docks proper the Northumberland and Albert Edward Docks were the first constructed. Here a large output of coal took place, for the traffic line along which the steam coal from the Northumberland coalfield was diverted once terminated here. But the development of the river Blyth has altered all this, and the latter now exports what only rightly is her own. By systematic dredging and deepening of the river former dangers have been removed, for the harbour is now amply protected, and for over a mile the channel is more than 30 feet deep at low water ordinary spring tides. Instead of sandbank and shoal we have the fine Tyne Dock, Albert Edward Dock, and Smith's Dock, while rows of vessels can lie four or five deep on either side and still leave a fairway for traffic.

But to obtain the best impression of the Tyne's

The Tyne Ports, looking west

importance one must make the journey from the piers to Newcastle. The Fish Quay, warehouses, repairing yards, and docks line the banks of North and South Shields. Then we pass the Coble Dene or Albert Edward and Northumberland Docks, with their stacks of timber from the Baltic. On the opposite bank we can see where Bede lived at Jarrow, and just below, a huge "slack" which is utilised for storing timber. Palmer's next claims attention, with its cranes, furnaces, rolling-mills, and vessels of all classes building and completed. In quick succession on the opposite bank we pass the Wallsend Slipway, the North Eastern Marine Engineering Company, Swan Hunter and Wigham Richardson, and Parsons'—all names great in the world of ships. Here we see a Cunarder on the stocks, there a Dreadnought approaching completion, here the steam turbine was made an actuality, there a pontoon is ready for South America, and so on for miles. These works are interspersed with minor yards, too many to claim special attention here, but each adding its quota to the Tyne's output and, in addition, chemical, copper-smelting and lead-refining works crowd one upon the other.

The black pall of smoke, the sunset peeping through as we journey westward, the incessant clang of hammer, the whirr of automatic caulker and riveter continuing for miles, combine to form a picture of the Tyne baffling adequate description. It is truly a mighty river, the like of which cannot be seen anywhere else the world through. And then, as if to remind one of the originators of it all, we see Stephenson's magnificent bridge spanning the river at Newcastle. Finally, we come to a succession of factory upon factory, engine shops, shipyards, ordnance works, and foundries by the mile. It is Elswick.

This, then, is "the Coaly Tyne," the evolution of 60 years, in itself a fitting monument to the greatness of the sons who made it.

## 13. Industries.

Of the industries of Tyneside and the development of the Tyne ports we have just treated, while the importance of the agricultural industry in the county has been shown in the chapter upon agriculture. Of the coal-mining industry we shall speak presently. In 1846 the shipment of coal from the Tyne ports was 3,265,334 tons, in 1894 it had reached 12,155,665 tons, and in 1911 it was no less than 20,185,343 tons. The shipping industry, employing so many of the population, is confined to the Tyne ports, Blyth, Amble, and Berwick. Here through a fine system of docks and staithes colliers are loaded and trimmed ready for sea in a few hours. Closely inter-related with this are the shipbuilding and general engineering industries, which employ the bulk of the people on Tyneside. These people are all engaged in the host of multifarious duties incident upon the construction of big ships, their engines, and, in part, of their cargoes. Thus there are thousands of riveters, caulkers, platers, fitters, engineers, pattern-makers, founders, brass-finishers and turners, each adding his quota to the "mammon of industrialism." The best way of obtaining an adequate idea of the number of men employed in this way is to watch the exodus of the workmen at Wallsend or Elswick at the close of the day. In shipbuilding the Tyne ranks second in Britain (the Clyde ranking first), her production in 1911 being 436,466 tons. Blyth produced 9258 tons in 1911. The importation and working of timber for ships and collieries is not inconsiderable and employs many men, and also allied to the shipbuilding industry is the manufacture of both wire and manila ropes at Willington.

It would be difficult to give a complete list of the multifarious trades and industries of our county. The

**Coal Staithes on the Blyth**

construction, fitting, and bending of copper steam-pipes have demanded special smelting works and foundries. Steel plates are made at Newburn. Lead-smelting and the extraction of silver is important at Wallsend. The glass industry was established at Newcastle by foreigners who sought refuge in the county about the reign of Elizabeth, and it soon spread along the banks of the

Tyne. The manufacture of anti-corrosive and anti-fouling compositions specially adapted for painting ships' bottoms affords employment in the Tyneside area. The chemical industry dates back to 1850, and about that time the manufacture of plate and bottle-glass was important and was also developed near Seaton Sluice, but to-day slag-heaps are the sole records of this industry. Pottery-making, the manufacture of glazed earthenware, and the production of fine bricks all afford employment along Tyneside. These have no doubt been encouraged by the huge deposits of alluvial clays, loams, boulder clays, and the fine seat-earths of the coal seams. The making of glass for resisting sudden changes of temperature or silica-ware is the latest introduction. Fishing and the concomitant curing and canning, fish-oil, and guano industries, are confined to the seaside places.

In the twelfth century Warkworth was noted for its salt-pans, and in the thirteenth century the salt industry had centralised around Blyth. In the fifteenth century it was almost the staple industry of North and South Shields. The tanning of hides and of nets dates back to the thirteenth century, and although the former is no longer practised, the latter still flourishes.

## 14. Mines and Minerals.

Minerals constitute Northumberland's chief source of wealth, and include coal, iron, barium, zinc, lead, whinstone, freestone, limestone, fire clay, and brick clay.

Coal, however, is the chief Northumbrian mineral and the large extent of the area of the coal measures in the county may be realised from the geological map. Wherever large pits have been opened in the coalfield we find huge aggregations of people employed in winning the coal. In Northumberland in 1911 there were 58,300 persons engaged in mining in over 100 mines. Such places as Wallsend, Annitsford, Seghill, Backworth, Killingworth, West Moor, Ashington, Shilbottle, and Cowpen all owe their origin to the rapid stride made in this industry. Coal seams are frequently associated with fine beds of shales and "seat-earths," and the winning of the coal necessitates the removal of these shales, which are frequently utilised in making bricks of fine quality. Consequently the "pit villages," and in fact most of the towns distributed over the coalfield, are wildernesses of bricks and mortar. Many of these shales contain large quantities of nodules of clay iron-stone. These bands have been worked as an iron supply, but not to any great extent, near Bellingham.

How early coal was worked it is difficult to say, but a thirteenth century document records the conferment of a privilege upon the monks of Newminster Abbey by means of which they were enabled to build a road to Blyth shore for the conveyance of coal. The Blyth coalfield was worked as early as the fourteenth century, and as a commercial product coal dates from the end of that century. It was shipped to London in Queen Elizabeth's days, though it did not attain much importance till the reign of Charles I. In 1908, statistics show that 13,797,527 tons, of a value of nearly £6,000,000, were raised and in 1911 the output of Northumberland was 14,682,427 tons.

The table appended gives the chief coal seams worked in the county, and their importance may be gauged by the approximate thicknesses shown:—

| | Thickness in Feet | District |
|---|---|---|
| Five Quarter Seam | 4 | Cowpen |
| Three Quarter Seam | 2 | Cowpen and Gosforth |
| High Main ("Wallsend") | 6 | Tyneside |
| Grey Seam and Blake Seam | 8 | Seghill, Cramlington |
| Yard Coal | 2½—4 | |
| Bensham | 2—5 | Preston |
| Low Main (best Gas and Steam Coal) | 2—6 | |
| Plessey Seam | 2 | Morpeth |
| Beaumont Seam | 3 | Morpeth and Preston |
| Hodge Seam | 2½ | Elswick |
| Yard Seam | 3 | Wylam |
| Brockwell | 1—4 | Preston |

In the *Report of the Royal Commission on Coal Supplies* (1905) the workable coal still remaining in the true coal measures was estimated by Sir Lindsay Wood at some 2,567,923,000 tons, and the amount within the three-mile sea limit at 1,112,910,000 tons.

In the twelfth century lead was exported from Newcastle and was probably obtained from near Hexham, as the mines there were prosperous in the sixteenth and seventeenth centuries. In the Allendale district lead ore was raised in small quantity. The by-products of this industry are barium sulphate and carbonate, and these salts, which are used in the manufacture of white lead, are worked at Fallowfield and Settlingstones near Newbrough, where in 1904 some 6286 tons of the ore were raised. In the same year the output of lead for the county had diminished to 67 tons. The methods employed are similar to those of most lead mines, the ore being obtained through drifts.

Igneous rock is worked mainly for road metal, and quarries are found along the outcrop of the Whin Sill. Some is converted into "setts" for paving streets, some into curb-stones, but the major portion, either hand-broken or machine-crushed, is used for making tar-macadam. For this purpose, however, it has a formidable rival in the slag from furnaces, but in 1908, nevertheless, 229,029 tons were quarried. Similarly limestones and sandstones are quarried along the various outcrops in the county; the former for road metal and for flux in smelting and lime-burning, the latter chiefly for building and converting into grindstones. The output of limestone in 1908 was 59,331 tons, and of sandstone 110,575 tons. Of clays used there are fire-clays, which owing to their refractory nature are useful in making fire bricks for furnace linings (208,666 tons), and brick clay for building purposes (123,917 tons).

Brick-fields with their dismal surroundings and sombre colour are common on Tyneside near the large population centres. Whilst the surface effects of coal-mining are not very apparent, areas of subsidence do occur and the fields in such areas are frequently of an undulating character. Farms and gardens, too, situated near the chemical centres and in the neighbourhood of smouldering pit-heaps suffer much from the fumes and smoke, and the sheep put out to pasture are often quite black with grime. Glacial sands and gravels are worked in connection with the building-trade, though in the seaside localities they are generally obtained from the foreshore. Instances are known of mines ceasing to work certain seams of coal, and the result is the migration of people from such centres in quest of employment. Peat bogs are common in certain moorlands but, when worked, peat is only for local consumption.

## 15. Shipping and Trade.

Some general idea of the importance of the Tyne ports has already been given. It is the object of this chapter to analyse in some detail the trade of the Tyne and the other Northumbrian ports. The Tyne and the Blyth are pre-eminently coal-exporting rivers, the former exporting in 1911 the astonishing quantity of 20,543,683 tons of coal and coke, representing an increase of 2,188,598 tons over 1910. Of this, 15,231,028 tons were for foreign and colonial consumption. The Blyth exported in the same year 4,424,945 tons, or twice her output in 1891, and an increase of 244,774 tons since 1910.

A considerable proportion of the import trade of any country must of necessity be food for the people, and this is peculiarly the case in non-self-supporting countries like

Britain. These food-stuffs will obviously come from the nearest centres of supply, in our case the Baltic and Western Europe. Then there must be raw material to supply the industries already established. Some idea of the enormous trade of the Tyne can be gleaned from the following. In 1911 she imported 48,944 tons of colonial food-stuffs, 276,150 tons of raw fruits, 35,568 tons of ale, 1486 tons of spirituous liquors, 1,124,062 quarters of grain and feeding stuffs, 18,619 tons of flour, 11,460 head of cattle, sheep, and pigs, 59,777 tons of provisions (beef, butter, bacon, etc.) and 65,246 tons of vegetables. From these figures it may be realised to what extent we are indebted to foreign supply for the very necessities of life, and how large must be the number of ships employed in this trade alone.

In 1911 the Tyne ports ranked third in the British Isles, 13,841 ships arriving with a net tonnage of 11,802,365 tons, and 13,821 ships leaving with net tonnage of 11,960,422 tons.

To supply her vast industries the Tyne is a great foreign buyer. In 1911 she imported iron goods to the extent of 38,838 tons, steel goods 43,282 tons, lead for manufacturing and de-silverising 49,405 tons, manures 34,235 tons, oils, varnishes, etc., used chiefly in shipping 23,970 tons, wood for shipbuilding and mining (pit-props, etc.) 328,289 loads, pulp for paper-making 17,161 tons, cement 36,421 tons, and clay, bricks, and tiles 15,051 tons. The chemical and fishing industries demanded 19,588 tons of salt, and 4176 tons of Norwegian ice were imported mainly for the fishing and frozen-meat industries. The chemical industry moreover necessitated 63,293 tons of sulphur ore, smelting claimed 29,836 tons of silver sand, 824,574 tons of iron ore, and 101,076 tons of burnt sulphur ore. The foundries took some 60,000 odd tons of chalk, loam, etc.

The Blyth imports consist in the main of mining and other timber 31,421 loads. The expansion of colliery villages brought in 1600 tons of cement and 1280 tons of gravel. Blyth imported 23,140 crans of herrings in 1911. The gross tonnage of all vessels, British and foreign, using the port of Blyth in 1911 was 2,167,937 tons, nearly double that of 1891.

The export trade of the Tyne consists primarily of coal, coke, grindstones, chemicals, manufactured iron and steel, and clay goods. The chemical industries are responsible for 132,747 tons of export. Of clay goods and fire-bricks there is an output of 112,597 tons, of iron goods 73,972 tons, of steel goods 28,854 tons, of paint 11,437 tons, of purple iron ore 81,626 tons, and of zinc and lead ores 10,893 tons. In addition to coal, the port of Blyth exported 4872 boxes of fresh herrings and 17,941 barrels of pickled herrings, chiefly to Germany and Russia. She also had a large output of casks.

There are regular lines of steamers leaving the Tyne for Antwerp, Hamburg, Rotterdam, London, Aberdeen, Leith, Hull, and in the summer months for Norway.

The principal import of Berwick-upon-Tweed is sawn timber. The chief exports are coal, grain, salmon, and manure. Amble is a potential outlet for the northern part of the coalfield.

Of extinct ports the most important is the silted-up Seaton Sluice, where there were formerly flourishing glass, salt, and copperas industries. Vessels of 300 tons could enter this artificial harbour in 1763.

## 16. Fisheries and Fishing Stations.

The fishing industry is one of the most important of Northumbrian occupations, and employs a large percentage of the maritime community extending from Berwick to North Shields. It is undoubtedly of great antiquity, dating as far back as 495 A.D. for East Britain, but there has been a surprising increase in the catches and exports in the last century owing to a variety of causes.

Fishing as an industry is necessarily dependent on good harbours, but it is of the first importance that these should be (1) within easy reach by rail of large industrial centres, (2) within a short distance of the fishing grounds, (3) where the people are seamen and fishermen born and bred.

In the early days of sailing-boats the demands made upon ports were not so great as they are to-day. Now the newer methods of fishing call for easy supplies of coal for trawlers, or oil or petrol for drifters, ice and salt for packing and preserving the catch, suitable sheds, quays, and docks for the auction and retail of the fish, besides rope-works, net-tanning tanks, paint-works, kippering-works, guano-factories for the utilisation of the waste products, canning-factories, and an express railway service for the carriage of the produce to the great centres of population. Naturally those districts which have provided the above facilities have made great strides, and the concomitant industries are quite as important in some areas as the fishing industry which first caused them to arise. There are three great branches of the latter—that of herrings, of white fish, and of shell-fish. The regularity of the migrations of the shoals of herring and mackerel has been known for a long time, and the greatest catches of the former take place in the summer months, about July and August. They are made by steam-trawlers or by ordinary sailing boats fitted with petrol-driven motors. These boats make for certain banks or shallows which by experience are known to be the resort of the fish at certain seasons. Fish life is controlled by many factors, but chiefly that of food. Cold and salt affect the plants and animals in the sea. The North Sea is shallow and is warmed by the North Atlantic Drift blown to these shores. Shallow waters around continents, under such conditions, favour the growth of sea-plants and the minute creatures upon which fish feed, so that we expect to find fisheries in temperate seas of moderate salinity in moderately high latitudes, which is precisely what occurs.

The fishing industry continues all through the year, but in Northumberland it is at its height from July to September, and at this time the ports swarm with strange boats and strange dialects are heard. A migratory community appears with the shoal, such as packers, cleaners, salesmen, and clerks, most of whom come from Scotland. By far the greater part of Northumberland's fish-supply is obtained by trawling, and consequently we find the largest fish markets at North Shields and Blyth. Most of the other fishing is either carried on by means of cobles or sailing vessels, steam-drifters, or motor-boats. It is a pretty sight to see the fishing fleet leaving these ports in the summer months, and particularly interesting is the race back to reach the market first. The smaller centres of the fishing industry stretch from Berwick to North Shields, and are Holy Island, North Sunderland, Beadnell, Newton, Craster, Boulmer, Alnmouth, Amble, Hauxley, Cresswell, Newbiggin, and Cullercoats. A compilation of the statistics of all the fishing stations in the county for the year ending September 1912 shows that there were 467,328 cwt. of white fish (cod, haddock, whiting, ling, halibut, plaice, dab, soles, catfish, skate, coal-fish, and conger eels) landed, of a total value of £352,868. North Shields, Blyth, and Newbiggin had the biggest catches. With reference to the herring industry the total catch was 370,348 cwt., valued at £136,085, and in this North Shields, Blyth, and Berwick-with-Spittal occupy the first places. Mackerel fishing is chiefly confined to North Shields and Blyth, 7819 cwt., worth £4941, being landed. The smaller stations all along the coast afford very pretty scenes with their picturesque bays and cobles, and their quaint inhabitants. It is in shell-fish that many of these specialise, and particularly in crabs, lobsters, and periwinkles. The large total of 1,183,972 crabs were caught in the year

Fish Quay, North Shields

under consideration, valued at £8412, and the ports which contributed most were (1) Beadnell, (2) Craster and Cullercoats, and (3) Holy Island and North Sunderland. The prawn or Norway lobster (*Nephrops norvegicus*) is landed from trawlers at North Shields, 13,732 cwt. valued at £5663 arriving in the past year.

Of lobsters proper, there were 47,899 caught, valued at £2542, and this catch, it is interesting to note, is distributed along the coast, the greatest number being claimed by Holy Island, Hauxley, and Beadnell. Only six stations engage in the periwinkle trade, 1885 cwt. being obtained, worth £453, and of these Holy Island has the greatest output. The gross value of the fishing of all kinds (exclusive of salmon) at all these ports for the past year was £510,964. This value was exceeded in 1909 and decreased to 1911. The mean, however, of a series of years shows that the total value is "fluctuatingly constant" about £511,000. Salmon fishing employs many hands in the summer months, and in the winter many of the workers go "into the pits" in the industrial area.

## 17. History of Northumberland.

Caesar landed in Kent about half a century before the birth of Christ (55 B.C.—54 B.C.), but it is not till the dawn of Christianity that we have any authentic British reports of our county. It was, however, destined to be an important factor in moulding the history of

Britain. Containing the most remarkable monuments of Roman greatness, the cradle of Christian influence, and later the home of the Percys, who figure on nearly every page of mediaeval history, it is not to be wondered at that Northumberland is of deep interest to the historian. By the orders of Vespasian, Julius Agricola came to Britain in 78 A.D. as his representative. Four years were spent in subduing Lower Britain, when his attention was directed to Caledonia. In 80 A.D. he had crossed what is now Northumberland and entered the Tay valley. After defeating Galgacus in 84 A.D. in the Grampians, he was withdrawn by Domitian. It is probable that he and his generals fashioned the Roman roads in Northumberland and passed on the merits of the Tyne Gap as a strategic point to his successors. For 36 years after this there is little record of Northumbria till the advent of Hadrian in 120 A.D. Much that Agricola had attempted this famous general had to repeat. It is to this period of English History that the Wall of Hadrian or the Picts' Wall, as it is called, is due. He left Aulus Platorius Nepos to finish it. This famous rampart, which is more fully described in the chapter on Antiquities, extends from the Tyne to the Solway, but the part where it is best seen lies between Haltwhistle and Chollerford and is shown on the map on page 107. This map brings out the geographical relationship of the Wall to the high contour of the Whin Sill, and shows six of the chief camps, besides a Roman bridge and "street."

Numerous are the relics testifying to the grandeur

and luxury of the life of the garrisons holding the Wall. In the fourth century Roman power declined and about 411 A.D. her dominion had ceased, the Emperor Honorius having granted the Britons independence. Northumberland, now unguarded, and weakened by Roman dominance, was easily open to the raids of the Picts and Scots, as well as the inroads of the Teutonic peoples from Low Germany and Scandinavia.

The Angles, one of three ravaging tribes, settled largely in North Britain. Augustine first visited England at this time, being sent by Gregory the Great. From the Forth to the Tyne extended the kingdom of Bernicia, and that of Deira from the Tyne to the Humber. The history of England as the land of the English and not of Jutes, Saxons, and Britons, commences with the foundation of the kingdom of Bernicia. Ella, King of Deira, conquered Bernicia, and Ethelfrith united Bernicia and Deira under one sceptre. Ethelfrith, King of Northumbria, defeated the British at Chester in 613 A.D., and the Northumbrians overran Westmorland at this time. About the year 627 A.D., Edwin of Northumbria became Christian through the influence of Paulinus. Many parts of Northumberland are associated with the name of Paulinus, particularly Pallinsburn and the Lady Well at Holystone in the vale of the Coquet, where Paulinus baptised three thousand Northumbrians. Bamburgh Castle, a great rocky fortress built in an impregnable position on the Great Whin Sill, occurs often in the history of this period. Here Ida commenced his reign in 547 A.D. It was known to the Celts as Dinguardi, and figures in the Arthurian Legend as Joyous Garde—"King Ida's castle huge and square." Ethelfrith, who was Ida's grand-

Holy Well of Paulinus

son, named it after his second wife Bebba—Bebba-burgh—and Bebba held it against Edwin of Deira. Oswald—

the great King—reigned from 634 to 642 A.D., and fought one of the most important battles in English history at Hevenfelth (Heavenfield or St Oswalds), for it finally crushed the Celtic power, and there were afterwards no more aggressive movements by the Britons against the English.

Heavenfield

English sacred literature may be said to begin in this period (seventh to ninth centuries). Caedmon is rightly termed the Father of English literature. The period is also associated with many famous Northumbrian names: Edwin, Oswald, Oswy, Aidan, Wilfrid (633), Cuthbert (637), Bede (673), and Alcuin (735).

In 635, St Aidan founded the monastery of Lindisfarne or Holy Island, where St Cuthbert succeeded him. Great advances were made in the county in the period of Saxon domination, but desolation and destruction attended the Danish incursions of the ninth century. The details of this period are very obscure, but border feuds, common for centuries, were numerous at this time. Bede died in 735. In 750 Offa laid claim unsuccessfully to the Northumbrian throne. In 774 Bamburgh served as temporary refuge to Alcred, King of Northumbria, before his banishment to Pictland. The Danes destroyed the long-established churches at Tynemouth and Lindisfarne in 912, but Bamburgh still held her own, until Athelstan conquered her in 924. The title of King of Northumbria was merely a name for about 25 years, Bamburgh eventually becoming the seat of the succeeding line of Earls. Being sacked in 993 by the Danes under Justin and Guthmund, it was subsequently used by the enfeebled Earl Waltheof, who sought refuge here in 999 during the invasion of Malcolm, the son of Kenneth.

With the coming of the Normans (1066), Alnwick and its lands were granted by William to his standard-bearer, Gilbert Tyson, and it was not till 1309 that the Norman Percys from Yorkshire came into possession, as Wardens of the Marches. Robert of Normandy had been deputed by his father, William the Conqueror, to build a castle at Newcastle in 1080. William Rufus, his brother, commenced to build in its place a new castle, but the present keep was not completed till 1177. The adjoining Black Gate Museum is an entrance to the court-

The Castle, Newcastle

yard of this castle dating from the same period. In 1095 Robert of Mowbray, the third Norman Earl of Northumberland, refused to appear at the Court of William Rufus. William advanced into the county and captured Tynemouth, Morpeth, and Newcastle. William the Lion was captured at Alnwick, near the Lion Bridge, in 1174. Otterburn, immortalised in the Ballad of Chevy Chase, was fought in the reign of Richard II (1388). The Scots under James Douglas harried Northumberland and Durham and affronted Harry Hotspur, Lord Percy, at Newcastle. He pursued them to Otterburn and was defeated, though Douglas was slain. The Battle or Percy Cross marks the spot where Douglas fell. Froissart says "this was the hardest, and most obstinate battle that was ever fought." Warkworth Castle was the home of Harry Hotspur in the days of Henry IV, and the Hotspur Bar is a landmark in Alnwick. Shakespeare makes Prince Henry, afterwards Henry V, say of Hotspur:—

> "I am not yet of Percy's mind, the Hotspur of the North; he that kills me some six or seven dozen of Scots at a breakfast, washes his hands, and says to his wife, 'Fie upon this quiet life!—I want work'." *Henry IV*, Pt. I, Act II, Sc. iv.

The centre of strife of the Wars of the Roses in 1464 lay in Northumberland. The battle of Hexham took place on the east side of the Devil's Water near Dipton Wood. The Lancastrians were defeated and the Duke of Somerset taken and executed at Hexham. Hedgeley Moor also witnessed a struggle at this time. Again in 1513 fierce fighting took place, this time at Flodden Field near Ford Castle, eight miles from Wooler,

The Battle or Percy Cross, Otterburn

and near the Scottish border. Here James IV and the flower of the Scottish nobility were slain. Tranquillity reigned till the time of Charles I, when Newcastle sided with him in the Civil War, but in 1640, after the battle of Newburn, the town was surrendered to the Scots.

The last great rising in the North was in 1715, in favour of James Stuart the "Old Pretender," the son of James II. General Wade encamped upon the Town Moor at Newcastle in the later rising of Charles Edward —the "Young Pretender"—in the year 1745. Thus it was that modern Northumberland issued out of the semi-barbaric past through long vicissitudes of bloodshed, ravage, and fire; watchful, dogged, and persevering. These characteristics were handed on to her sons, who later were to gain glory in the greatest exploits at sea, in war, and in the more peaceful path of invention.

## 18. Antiquities—(*a*) Prehistoric.

The history of Britain is written to no small extent in her treasures of antiquity, but the earlier chapters are necessarily more difficult to understand than others because the relics of our more remote ancestors are limited both in number and character. We shall deal now with these obscure chapters of which we have no written records.

Of Prehistoric men of the Palaeolithic Age we have no definite traces in Northumberland. Their successors however, the Neolithic men, buried their dead in "barrows," and it is in these that we not infrequently find relics of them. They comprise flint arrow-heads and knives, necklaces of jet or shale beads, stone celts, perforated axe-heads, and sometimes incised rocks. The stone implements are usually found in the fells on moors or in peat bogs, and it is here that these people were driven by later invaders. Celts have been found at Doddington, Harbottle, Halton Chesters, Bellingham, and Ilderton. They vary in dimensions, being from 5 in. to 6 in. long, and over 2 in. broad, and they were no doubt fastened to wooden or bone shafts. The next development, the perforated celt, has been found at Shilbottle, Alnwick, Haydon Bridge, and Seghill. A cist was discovered at Seghill (1866) containing one of these made of quartzite and fashioned concavely, and another of basalt is recorded from Twizel.

Curious stone balls, 2¼ in. in diameter, presumably used in games, have been discovered on Corbridge Fell. Flint flakes have been found at Chollerford, Amble, and Ford, and numerous cists have come to light near Hawkshill township. The Spindleston Crags have afforded fine sepulchral pottery of the Bronze Age, some of which is preserved in the British Museum. In the Newstead township in Rayheugh Moor three large cairns were discovered, which yielded cists, skeletons, cinerary urns with burnt bones, and drinking cups, which had been ornamented by scratching with notched bones. Similar remains are recorded from Lesbury and Longhoughton. The Lesbury drinking cup was beautifully ornamented with five horizontal zones of dotted lines. Food-vessels of the Bronze Age ornamented with marks made by twisted thong were found at Hazon, Low Buston, and

**Menhir at Swinburn**

also in the township of Bolton. A cairn at Amble made of pebbles from the seashore afforded a variety of urns of similar design. Many and varied have been the relics revealed at Birtley, including three distinct hut circles, cairns, cists, urns, ashes, pottery, querns and cup-marked stones. Bronze weapons have also been found. At Newham Lough a fine socketed bronze celt was obtained, a bronze spear-head was found at Elford, and in a cairn at North Charlton there were four cists, one containing a skeleton, on the breast of which lay a bronze knife-dagger, besides charred remains and an ornamented urn. Fine axe-heads have been found at Tinely, Boulmer, and Shilbottle, the former being of polished green felsite, and the latter, found in Long Ridge field, being of green-stone.

A notable menhir or standing stone stood in the park of Swinburn, measuring 11 feet in height by 3 in width. Here is the site of a group of sepulchral barrows, in which were five cistvaens. Of incised stones the chief occur at Old Bewick and Doddington, the markings being chiefly spirals and ornamented circles. A fine camp circle forms part of the antiquities of Gunner Park.

## 19. Antiquities—(*b*) Roman. The Roman Wall.

The Roman Wall, to which several references have already been made, is the chief of Northumberland's antiquities. Its course can be traced on the map extending from Bowness on the Solway to Wallsend on the Tyne, a distance of nearly 74 miles. This wall, "The Pict's

Wall," as it is sometimes called, was a distinct military defence, and in no way can we look upon it as an attempt to delimit the Roman frontier in North Britain, for, were that so, it is obvious that the camps disposed along it would not have afforded ingress from the north, nor would Roman camps have existed north of it as they undoubtedly do. The various parts of the defence are illustrated in the diagram, p. 108 (which is after that by Bruce), and their disposition can be best seen near Borcovicus, and also at Carr Hill near Halton Shiels. Speaking generally, the fortifications consist of a stone wall with a ditch on the north side, roads, mile-castles, turrets or watch towers, stations or camps, and an earth wall or vallum to the south of the stone wall consisting of several ramparts and a fosse. The Wall is characterised by the directness of its line across country, only deviating from this to take advantage of the strategic strength of a particular eminence. The original height of the Wall is unknown. The Venerable Bede gives its dimensions as 12 feet in height and 8 feet in breadth, though these measurements must be looked upon as those of a particular locality. However, from the average of investigations it is reasonable to suppose that its height was originally about 20 feet, though to-day it varies up to 9 feet only. Probably the width was about 8 feet, but this also varies, which, perhaps, may be accounted for by the assumption that the respective centurions, disposed at intervals along the Wall and superintending its erection, built their sections to meet what were deemed to be local requirements, and that these were subsequently united. A deep fosse 36 to

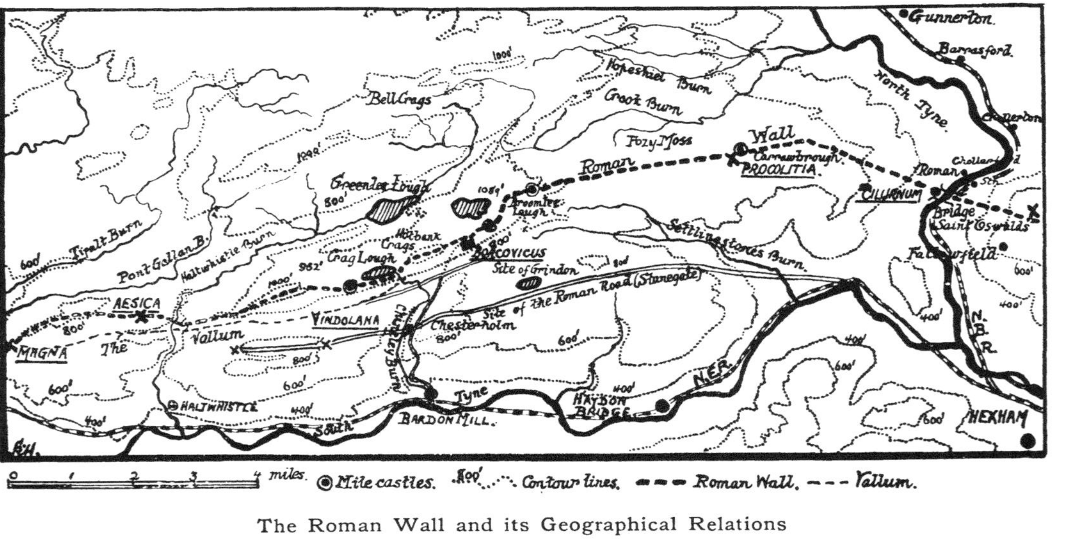


The Roman Wall and its Geographical Relations

40 feet wide, and 12 to 15 feet deep, runs along the whole of the north side of the Wall.

Starting at Wallsend (Segedunum) the Wall strikes for the site of Newcastle (Pons Aelii), where it was met by the Roman way from Chester-le-Street. It must here be pointed out that Newcastle does not owe its site to its position on the Wall, as is sometimes stated, but as in the case of many towns the contributory causes have been mainly geographical. Here also Newcastle exemplifies that type of city arising at the intersection of two lines of weakness—the then easily fordable Tyne, and the

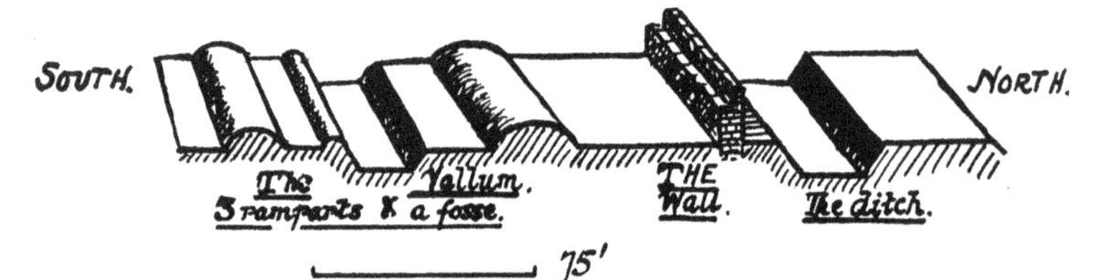


Diagram of the Roman Wall

Team Valley-Ouseburn gap—as is readily appreciated from the map. The Wall leaves Newcastle and passes Benwell (Condercum), the third station, two miles from Newcastle. Continuing through Walbottle it persists in its direction to Vindobala, the fourth station, just past the mile-castle of Rutchester. It thence proceeds to Harlow Hill and Halton Shiels, bringing us to Hunnum, 7¼ miles from Rutchester and just east of the intersection of the Watling Street, which has crossed the Tyne *via* Corstopitum (Corbridge). The same line is maintained from Hunnum to Procolitia at Carrawburgh, having crossed the North Tyne by the famous bridge below Chollerford and bisected the camp at the Chesters (Cilurnum), the second largest station on the Wall. Procolitia is 3½ miles westward from Cilurnum, and here the course of the Wall is diverted to south-west by west to occupy the fine crag-lands of the Whin Sill at Borcovicus, with the freshwater fish and fowl supplies of the Northumbrian lakes to the north. Borcovicus or the Housesteads, five miles from Carrawburgh, is an important "type station" which, like the Chesters, we shall consider shortly.

Chesterholm (Vindolana), about half-way between Bardon Mill and Housesteads on the Stanegate, is the site of a station remote from the Wall but on the main Roman road. This road deserves special notice, for here can be seen a fine Roman milestone *in situ*. A quarter of a mile to the west of Borcovicus is a typical mile-castle—a building of a class so called because they were placed at intervals of a Roman mile, and were obviously meant to accommodate the garrisons holding the Wall. The track zigzags now till the "Nine Nicks of Thirlwall" are reached. Continuing from the mile-castle we come to the camp of Aesica or Great Chesters, the tenth station on the Wall. Beyond the Nine Nicks of Thirlwall lies the camp of Magna or Caervoran, the last in the county, and situated 2½ miles from Aesica. At intervals between the mile-castles, turrets or watch-towers were constructed, no doubt serving the same purposes as the sentry-boxes of to-day.

Having thus outlined the general trend and features of the Wall, let us now examine some of its principal

Roman Milestone

camps. The best for this purpose are Cilurnum (The Chesters), Borcovicus (Housesteads), and Corstopitum

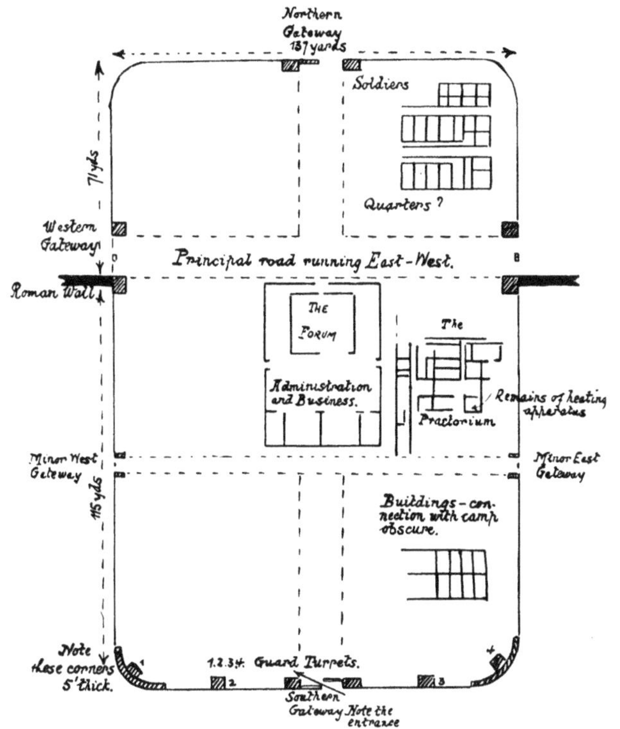


Rough diagram showing disposition of chief points of interest.
- CILURNUM -

(Corbridge). Here, thanks in no small measure to the researches and enterprise of Mr John Clayton, we can gather something of the life of the men who garrisoned this wonderful structure. Cilurnum embraces a camp, a villa, a bridge, and a fine museum, which deserve especial study. Our rough diagram shows the relation of the site of the camp to the Roman Wall as well as its approximate dimensions and the disposition of its chief buildings. It is roughly about $5\frac{1}{4}$ acres in area, and is the second

Roman Street, Cilurnum

largest camp *per lineam Valli*. In general the shape is like that of the other camps, although for some reason its orientation is different. Geographically speaking the site, like that of Borcovicus, is admirable, and judiciously chosen. It lies about 100 yards from the river, thus affording a good water and fish supply, the former especially important to the bath-loving Romans. Cilurnum however differs from Borcovicus, inasmuch as the Wall forms no part of it but joins the camp at the south gate-posts of the east and west gateways. It has six gateways, whereas the other camps generally have four. The eastern gateway led to the bridge and is much finer than the rest. They still show the ruts on the threshold worn by the

The Forum, Cilurnum

coming and going of chariots and traffic, the gutters for carrying away rain, the sockets on which the gates worked, and the central threshold stones against which they closed. The streets are at right angles as in the other camps. The central buildings, doubtless a combination of forum and business offices, are well preserved, and the shops were probably arranged around this structure.

Justice was also dispensed here, and the treasury was situated in the south central building of this group. The praetorium with its bath and heating chambers formed the residence of the Governor. When opened out in 1843 the flues of the heating chamber still contained soot, and the bath was coated with a red cement. Here it was that the North Tyne river-god was found, now preserved in the museum. Coins of a variety of periods, gems, and styli, used in the graving of tablets, all occurred in this bath-room. Remains of two other groups of buildings exist in the camp, one to the north and the other to the south of the praetorium. The latter are difficult to make out, but the northern group with its fine guttered street and heating apparatus (hypocaust) most probably formed the soldiers' quarters. The granaries lay just beyond the forum. From the position of the camp it seems that the south side was the weakest. It was readily accessible from the easily-forded river, whose wooded banks would afford good cover for invaders, and therefore we are not surprised to find evidence of strong fortification at this part in the four turrets, which form a striking feature.

Leaving the camp behind, an interesting though obscure group of buildings in a fine state of preservation is reached. They are contiguous with the river above the flood-limit and are known as "The Roman Villa." Opinion is divided as regards their object, some attributing to them the dignity of the Governor's residence, and others believing they were a combination of bath and temple. The most interesting features are the seven arches and the hypocaust. Here was found an altar dedicated to Fortune.

Quite near is the Roman Bridge, which is recorded as far back as 1599. The rough diagram is after one made for Mr John Clayton's paper and is almost self-

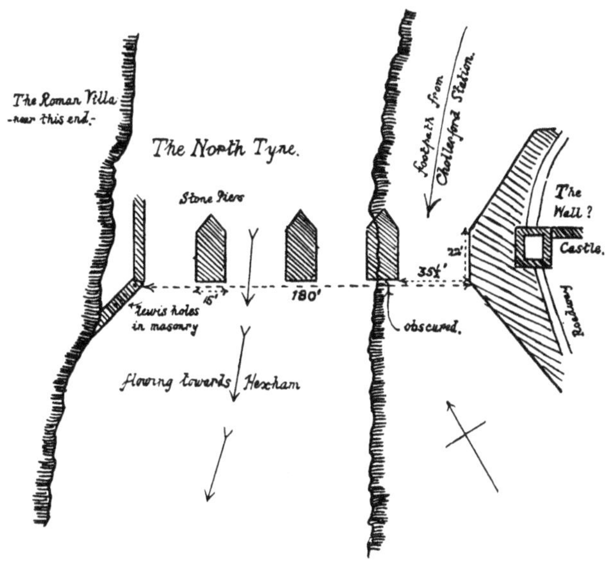


The Roman Bridge at Chollerford

explanatory. There is abundant evidence here of the high degree of engineering skill attained by the builders. The observer is struck particularly by the way in which the fine, faced stones are joined together with iron ties embedded in lead. "Lewis holes" also are still to be seen, and many of the blocks are of large dimensions. The bridge is ascribed to Hadrian, but it is more than likely that an earlier one existed here. Hadrian's bridge was subsequently repaired by Severus.

The readily-forded river, the pleasant situation of the camp shut in on one side by the Limestone Bank towards Walwick, and on the other by the Brunton Bank towards St Oswald's, the plentiful supply of water, fish, and fowl, the abundant fuel, and the huge outcrops of fine building-stone at Brunton, all show the work of one who built "not for a day but for all time." It is not wonderful that a vast store of antiquarian treasure has been found, and there is abundant evidence that here, at least, luxury reigned.

The dreary site of Borcovicus contrasts strongly with this camp. It is truly a city of the dead, impressive in its intense solitude. A visit must be made to Borcovicus in the depth of winter fully to appreciate the rigours of its position. It resembles Cilurnum in the outline of its parallelogram. There are four gateways and two streets joining them; the north and south or the *Via Principalis*, and the east and west or Pretorian Street. Here, in the walled-up gateways, is evidence of the waning glory of Rome. It was by making a diagonal entrance into the camp (see diagram Cilurnum) that the weakened garrisons sought security.

Samian pottery, as it is called, a red ware made in Gaul and largely used by the Romans, was found here abundantly, together with coins and other relics. It was near the Decuman or south gateway that the famous Mithraic cave-altars and figures were unearthed, all

Roman Wall looking East towards Housesteads

bearing witness to the worship of Mihr or Mithras, the Persian sun-god, Apollo, which had been introduced into Europe in the time of Julius Caesar, and which was usually accompanied by human sacrifices. This god is associated in the relics with the twelve signs of the Zodiac. The figure of Victory, so great a favourite with the Tungrian Infantry, was also found here.

The Pretorium, Borcovicus

In excavating the mile-castles many inscriptions were found referring to Aulus Platorius Nepos, Hadrian's legate.

Corstopitum, near Corbridge, has been the scene of extremely interesting researches from 1906 to 1911. It may be taken as typical of a Romano British station away from the Wall. Here Watling Street crossed the

Tyne by a bridge and joined this famous settlement, originally founded by Agricola in 79 A.D. It revived in the reign of Hadrian, and subsequently rose to great importance as a distributing centre. Recent research places its area at about 30 acres. It was not a fort and relied upon the Wall for protection. There is evidence

**Fountain and Granary, Corstopitum**

that pottery, tools, and arms were manufactured here. The streets were planned as elsewhere north-south and east-west, and the most important buildings uncovered are the Forum and the east and west granaries. The Forum, probably in this case a huge military depot, covers an area of 220 feet square and contains a central court about 170 feet square, and a series of buildings having access to the streets surrounded it. The granaries, both of enormous size, are situated to the west of the Forum, and are fine buttressed buildings of different periods. The western is the older and longer of the two, and is about 93 by 24 feet, the other being seven feet shorter and a foot wider. Just outside the eastern granary, at the south corner, is a large fountain and trough supplying the water of the town. The masonry of the west wall is considered to be the finest in Britain.

The treasures of this camp are mainly exhibited in a temporary museum and include carved stone monuments, the Corbridge Lion, which was used as a fountain supplying a water-tank and is of very primitive workmanship, the fine altar to Jupiter Dolichenus, the native Castor ware from Northampton, a fine collection of Samian ware, home-manufactured *mortaria*, cooking-jars, wine-jars from Spain and France, window glass, candlesticks, crowbars, pick-axes, spear-heads, ballista-balls, and caltrops.

Chief among the treasures found in 1911 is a repaired bronze jug, which contained a hoard of gold coins, 159 in number and of various ages from the time of Nero to Antoninus Pius. This find, as well as a prior one in 1908, are now housed with the National treasures in the British Museum.

In addition to this and the articles above mentioned, dice, a draughts-board, a fish-hook, finger-rings and pendants, bronze *fibulae* or brooches, bone needles and pins, etc., all bear witness to the life in this interesting Roman town.

Hexham Abbey is also singularly attractive to antiquaries, for in its crypt can be seen the use to which the dismantled wall has been put by the successors to the Romans. Fine toothed work and two inscriptions are of importance. On one of these can be observed the erasure of the name of Geta, brother of Caracalla, and son of Septimius Severus. After the younger brother had been murdered to further the ambition of the elder, his name was erased from every inscription throughout the Empire.

The following table brings out the disposition of troops located on the Wall and shows how important it was as a Roman defence. Six thousand men, who were drawn mainly from Western Europe, garrisoned it. One camp was garrisoned from France, three from Spain, two from Holland, and two from Belgium.

| Camp | Name of Place | Garrison |
|---|---|---|
| Segedunum | Wallsend | Lingones; Lorraine, *France.* 600 Gauls. |
| Pons Aelii | Newcastle | Cornovii (? Carnovii); N. Wales (?), 1 cohort. |
| Condercum | Benwell | Astures; *Spain*, 1 ala. |
| Vindobala | Heddon-on-the-Wall | Frisii; *Belgium*, 1 cohort. |
| Hunnum | Corbridge | Savinii (? Sabini); *Centra Italy*, 1 ala (? Savus, *Austria*). |
| Cilurnum | Chesters | Astures; *Spain*, 1 ala. |
| Procolitia | Carrawburgh | Batavi; *Holland*, 1 cohort. |
| Borcovicus | Bardon Mill | Tungri; *Belgium*, 1 cohort. |
| Aesica | Haltwhistle | Astures; *Spain*, 1 cohort. |
| Magna | Nine Nicks of Thirlwall | Dalmatae; *Austria*, 1 cohort. |
| Amboglanna | Birdoswald | Daci; *Balkan Peninsula*, 1 cohort. |

## 20. Architecture—(*a*) Ecclesiastical.

The county of Northumberland—"the cradle of Christianity in England"—is singularly rich in pre-Reformation churches and affords examples of every type of architecture from Saxon to Perpendicular. The chief periods of English church architecture are:—

Saxon or pre-Norman up to about 1040.
Norman or Romanesque—1040 to 1180.
Early English—1180 to 1300.
Decorated—1300 to the Black Death (1348).
Perpendicular—about 1360 to the dissolution of the monasteries in the middle of the sixteenth century.

No definite line of demarcation can be fixed between these successive styles. Transitions occur throughout, partaking of the essential characteristics of each, and these dates are only broadly correct. Saxon architecture was of a very simple nature. The Anglo-Saxon style is characterised by rough rubble walls, no buttresses, semi-circular and triangular-headed doorways and deeply-splayed windows inside and out, and square towers with what is termed "long and short work" at the corners, and the stone buildings of this period were modelled on the earlier wooden erections. St Wilfrid's Crypt at Hexham is the only surviving portion of St Wilfrid's church and is distinctive of this type. Many of the stones of this church, as at Warden, testify that it was fashioned of Roman masonry and both Roman cornices and capitals have been utilised. Fine examples of the single-light windows with double splays are found in St Andrew's at Bywell. Characteristic belfry windows of this period with semicircular-headed lights separated by a baluster-shaft, which supports the "through-stone," occur at Bywell St Andrew's and Ovingham. Further examples occur in the county, notably at St Andrew's, Corbridge, St Andrew's, Hexham, Bolam, and Heddon.

The great church builders were the Normans, who frequently demolished the Saxon churches, and replaced them by richer and more majestic types, represented by remnants at Seaton Delaval (interior), the ruin of St Peter's church Holy Island, St Mary's Holy Island, and the chapel in Newcastle Keep amongst others. The early Norman or Romanesque style of architecture was characterised by heavy walling, richly carved and deeply set doorways, semicircular vaulting, and short massive pillars supporting round arches. The capitals were large and square with the lower quoins rounded off. The windows were still small and narrow with one-splayed semicircular heads. The towers were usually short and massive and placed at the intersection of the nave and choir with the transepts.

These were succeeded by churches characterised by greater elaboration in vaulted stone roofs; in the choirs and transepts toothed ornament was beginning to be freely used. These churches illustrate the Transitional period,

Ovingham Church

exemplified in St Andrew's Newcastle, St Andrew's Hexham, St Peter's and St Paul's at Brinkburn, and St Mary's at Holy Island. The Priory at Brinkburn (1135) is one of the finest examples of the Transition from the Norman and belongs approximately to the same period as Hexham and part of Tynemouth, and shows the high

Seaton Delaval Church

standard of architectural skill then existing in the county. It is essentially Norman, intermixed however with distinct Early English.

From 1150 to 1200 the building became lighter, the arches pointed, and there was perfected the science of vaulting, by which the weight is brought upon piers and buttresses. This method of building, the "Gothic," originated from the endeavour to cover the widest and loftiest areas with the greatest economy of stone. The first English Gothic, called "Early English," from about 1180 to 1250, is characterised by slender piers (commonly of marble), lofty pointed vaults, long, narrow, lancet-headed windows and deep-cut mouldings. Later the

Lindisfarne Priory

windows became broader, divided up, and ornamented by patterns of tracery, while in the vault the ribs were multiplied. The greatest elegance of English Gothic was reached from 1260 to 1290, at which date English sculpture was at its highest, and art in painting, coloured glass making, and general craftsmanship at its zenith. This period is well shown in St Andrew's, Hexham.

The Nave, Hexham Abbey

After 1300 the structure of stone buildings began to be overlaid with ornament, the window tracery and vault ribs were of intricate patterns, the pinnacles and spires loaded with crocket and ornament. This later style is known as "Decorated," and the Newcastle churches of St Nicholas, St John, and St Andrew are partly representative of this period. It came to an end with the Black Death, which stopped all building for a time.

With the changed conditions of life the type of building changed. With curious uniformity and quickness the style called "Perpendicular"—which is unknown abroad—developed after 1360 in all parts of England and lasted with scarcely any change up to 1520. As its name implies, it is characterised by the perpendicular arrangement of the tracery and panels on walls and in windows, and it is also distinguished by the flattened arches and the square arrangement of the mouldings over them, by the elaborate vault-traceries (especially fan-vaulting), and by the use of flat roofs and towers without spires.

Canopied niches become more common at this period also, and the triforium is abandoned. This period is also partly represented by St Nicholas, St John's and St Andrew's, Newcastle; and by Tynemouth Priory.

Naturally, with so many ruined churches, abbeys, and cells, we find widespread traces of the chief monastic orders in the county. There are three Premonstratensian houses, notably Blanchland (1163) and Alnwick Abbey (1147), both dedicated to the Blessed Virgin Mary. The Carmelite Order, or the "White Friars," is represented

Tynemouth Priory

by Hulne Abbey, which was founded by a Northumberland crusader from Mount Carmel.

The Benedictine Order was the most widespread in this county, having cells at Amble, the Farne Islands, Coquet Island, the Priory church at Lindisfarne, the Holystone nunnery, and Tynemouth Priory. The Priory church of Lindisfarne is immortalised by Sir Walter Scott in *Marmion*. The whole of the life of a Benedictine can be traced in the remains at Lindisfarne, a typical building of the twelfth century.

The Augustinian Order came next to the Benedictine, having no less than five houses, and notably that at Hexham Priory (1113). The Dominican Order is represented by two houses, particularly the "Black Friars" monastery at Newcastle. The Cistercians held Newminster Abbey at Morpeth. The Knights Hospitallers had two centres in the county.

Of pre-Conquest remains there are several fine inscribed crosses. One was found near St Waleric's Chapel near the mouth of the Aln. This is partly inscribed in Runic and Roman characters. The Acca Cross from Hexham, now in the cathedral library at Durham, is also of pre-Conquest age. It is a magnificent type and is of Italian origin, the design being based upon the vine. Another was found at the Spital (Hexham), and Birtley church has also a sepulchral stone.

## 21. Architecture—(*b*) Military.

The medieval military architecture of Northumberland comprises castles, peel-towers, and walls. The necessity for protection from invasions along the borders was responsible for the fine examples of these found in the county.

The castles are placed in almost impregnable positions. Bamburgh Castle, perched on the Whin Sill, has an imposing outlook towards the sea, and it is difficult to say from which quarter it is best seen. William Rufus entered it by strategy after Earl Robert Mowbray had been decoyed to Newcastle, and thus it fell into the Norman's hands. It is entered through the barbican on the south-east, and through the sally-port gate on the north-west side. The Norman keep is inside the bailey, and the walls, which are of earlier date than the Norman keep at Newcastle, are from nine to eleven feet thick. There is an early Norman gateway on the ground floor, and from the absence of chimneys it appears that the object of this keep was, as usual, mainly one of defence in case of emergency. Coming next in grandeur is the Norman castle of Newcastle. The Black Gate, which was erected by Henry III, was the barbican entrance to the castle and led to the keep, which was finished in 1177. This is about 90 feet high, crested by battlements and towers, and is rectangular in shape, measuring some 56 by 62 feet. Here the walls are as much as 12 to 17 feet thick. On the east side of the ground floor is the

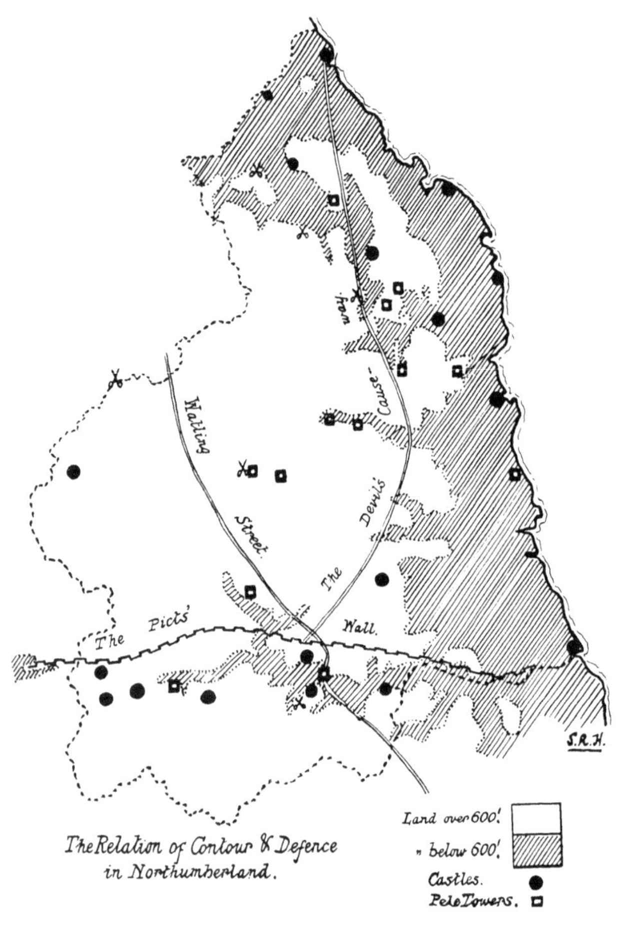


The Castles and Peel-Towers of Northumberland

*(To show their situation chiefly on or near the 600-ft. contour line)*

beautiful Norman chapel, part of which now forms a fine antiquarian museum.

Warkworth Castle, built in 1158, is of the Transitional Norman period, and stands on a hill on the south bank of the Coquet. The keep, some 80 feet square, dates from about 1435, and is one of the finest types in Europe. Four towers project from the sides and there is

Bamburgh Castle

also an observing tower : the chapel is in the eastern tower, and the windows show elegant tracery. In addition there are the entrance hall, oubliette, guard-room, grand staircase, and the great baronial hall. In fact this keep represents the transition from a pure fort to a fortified residence. It is the scene of a great part of Shakespeare's *Henry IV*.

Alnwick Castle is a fine fortified residence about 30 miles from the Border. It was the centre of incessant strife and naturally its historical importance is scarcely rivalled by any other in England. The names of Hotspur, Flodden, and Chevy Chase all recall Alnwick. The earliest parts of the first keep date from 1140. The castle itself was then as large as it is now and was in

Warkworth Castle

bad condition, owing to repeated onslaughts, when the first of the Percys obtained it. The fabric was nearly all rebuilt by them about 1309 to 1350. The present proud appearance of the castle dates from the time of Algernon, fourth Duke of Northumberland (1854–1864). The entrance is made on the west side by the fine barbican which was built about 1350. The figures on the barbican and battlements are very remarkable. The keep

Alnwick Castle

itself is in the inner bailey and had both moat and drawbridge. The entrance is through two octagonal towers and is characterised by the variety of the stone figures surmounting the battlements.

Dunstanburgh Castle, a ruin standing in a fine position on the crags of the Whin Sill (see p. 47), was the largest Northumbrian castle, but did not figure in history to any

Dunstanburgh Castle

great extent. This castle was rebuilt and transformed into a fortress in 1316 by Thomas, Earl of Lancaster. It suffered greatly in the Wars of the Roses.

Norham Castle is a fine grim-looking border-fortress in ruins, having a commanding position near the Tweed. The keep stands in the north-east corner of the castle site and is built of red sandstone, with walls from 12 to

15 feet thick. It dates from 1121 and was built to prevent Scottish invasions and border raids.

Prudhoe Castle stands in a picturesque position on the south bank of the Tyne and was one of the most celebrated strongholds of the north. It was founded in Henry II's reign and remained in the hands of the

Norham Castle

Umfravilles for 300 years. William the Lion failed to conquer it in 1174 after three days' siege. It was subsequent to this, on his retreat, that he was captured at Alnwick.

The Hotspur Tower at Alnwick (1450) is a relic of the time when the city was walled. Other walled towns were Newcastle and Berwick, the walls of the latter being still in good preservation and showing many fine gates or bars. The county is moreover prolific in peel-towers, many of which are in ruins and many incorporated in other buildings. Shilbottle Vicarage embodies one of these towers. An excellent example is to be seen in Whitton Tower, and it may be taken as typical of these guardians of ecclesiastical property.

Of modern fortresses Tynemouth Castle is the best instance. It was originally founded in the Norman period, and in 1644 was captured and partly destroyed by the Scots. It was repaired in the Civil War and again in 1782. The huge walls and the gateway are alone reminiscent of the early days.

## 22. Architecture—(*c*) Domestic.

Domestic architecture may be divided for our purpose into (1) famous seats and manor houses, (2) rural and town cottages. Further there are the fine municipal buildings so intimately connected with daily life in such industrial counties as Northumberland.

Of the first kind there are many remarkable and noteworthy examples, and this one would naturally expect in a county which has yielded such a large number of famous men. In many quiet villages where the modern hand has not been much at work we still find instances of the thatched cottage. But by far the greater number of houses are congregated in congested industrial areas laid out in a fashion which is anything but beautiful and aptly described as "huge wildernesses of bricks and mortar."

The finest type of domestic architecture is Alnwick Castle, of which we have already spoken. Of the remaining famous seats—and there are many—one domi-

Ford Castle

nant note runs through them all, namely, they have incorporated in them the remains of border fortresses or peel-towers. Of this type are Otterburn Tower, Eglingham Hall, Featherstone Castle, Bellister Castle, Unthank Hall, Whitton Tower, Ford Castle, Chipchase Castle, and Cresswell House.

Otterburn Tower is a fine modern castellated residence

incorporating a fourteenth century peel-tower. It was founded in the Tudor period. Eglingham Hall, the home of the Ogles, is a fine modern mansion also with a peel-tower. The building is associated with Cromwell. Featherstone Castle, originally founded in 1290, is similarly castellated and modernised. Bellister Castle and Blenkinsopp Castle are modern restorations of ancient buildings dating from the thirteenth and fourteenth centuries. Whitton Tower, half-a-mile from Rothbury, is one of the long line of peel-towers stretching from Warkworth to Hepple on the Coquet, and guarding the entrance to the lowland fertile plain. It is now the Rectory of Rothbury and is a typical peel with vaulted byre, barmkyn, and winding staircase to the summit. It dates from the fourteenth century and was founded by the Umfravilles. Ford Castle was originally built in 1287. James IV rested here prior to Flodden Field in 1513 and subsequently burnt it down. It is the earliest type in the North Country of the quadrangular fortress with distinct square towers at its corners, and it served as the model for the later type of Chillingham (1344). Only two massive towers now remain. It was restored (1764) by Sir John Hussey Delaval and modernised in 1865. Chillingham Castle, which possesses three towers of the thirteenth century, was designed by Inigo Jones in the Renaissance style and is a magnificent edifice. Odonel de Umfraville was the originator of Chipchase Castle, which was built in the thirteenth century and has been modernised. Langley Castle (1350), which has recently been restored, affords an excellent example of the "tower" type of house. It has some fine windows with Decorated tracery.

Cragside

Coming to more modern residences the finest examples are Cragside, the seat of Lord Armstrong,

Blagdon Hall, The Chesters, Beaufront Castle, Cresswell House, Howick Hall, and Falloden Hall. Cragside is the best instance of reclamation of what was practically a barren fellside. It is one of the finest houses in the county and stands in lovely grounds, a lasting memorial to the energy, enterprise, and perseverance of the first Lord Armstrong. Blagdon Hall is an imposing stone structure with a fine south front dating from the eighteenth century. Cresswell House, which includes the ancient peel-tower, represents the beautiful modern mansion (1821), and is built in the Italian style. Beaufront Castle dates originally from the Elizabethan period and is now a fine type of the modern castellated home. It is associated with the ill-starred Earl of Derwentwater. Falloden Hall, the seat of Sir Edward Grey, is a solid red-brick mansion.

Of picturesque cottages the finest are met with at Ford village and Etal. This is a distinctly fine area, and with its cosy, thatched houses and lovely gardens it has a delightful old-world appearance contrasting strongly with the dismal monotony of the pit-villages. Quaint lime-washed cottages are to be met with in the old fishing villages, notably at Cullercoats. The material used in the construction of houses, it should be noticed, varies with the type of rock. Where clay predominates the houses are of brick, but where sand or limestone occurs then the houses correspond. In the Lower Carboniferous areas the thin layers of freestone are utilised for roofing. The roofs of the oldest churches and houses of this type are fixed to the rafters by means of sheep-shanks, another instance of the utilisation of the material most readily accessible. George Stephenson's birthplace at Wylam is typical of numerous old country cottages, with their red-tiled roofs with thin sandstone "drip" and freestone walls.

## 23. Communications—Past and Present.

The chief factor which determined the lines of communication in our county was, as we have already seen, that of contour. Even before the days of the stage coach the passes in the Cheviot and Pennines were well known, and were used in travelling. Early people first migrated along the lowlands for security and ready supply of food, and tradition and usage determined with each subsequent civilisation that these tracks should develop into roads. They were undoubtedly altered considerably, but the general direction along lines of least resistance remained the same. Later, the advance of the Romans forced the Britons to occupy the mountain fastnesses, and thus we find their camps and barrows, and tracks leading thereto. Many of the Roman roads keep to these tracks. The pioneers in British road-building were the Romans, and though many have been abandoned yet several of our main roads are those which witnessed the passage of the Roman legions. The main roads constructed by the Romans were the Watling Street, the Devil's Causeway, the Stanegate, and the track of the Roman Wall (see p. 132). There is one remarkable feature about them all and that is their directness. They rarely deviate from a certain line and whilst using the river valleys when they can, they do not seek to avoid crag or mountain; in fact in some areas they appear to prefer the crest of a hill rather than a valley.

The Roman Wall we have considered. This was as much a road as a wall, and passed, as we have seen, right across the county from Wallsend *via* Newcastle, Heddon, St Oswald's, Walwick, Borcovicus, Greenhead, and Gilsland. Parallel to this and to the south, in the vicinity of Borcovicus, is the Stanegate, passing through Sunny Rigg, Chesterholm, and Settlingstones towards Newbrough. The chief roads of the county, however, were Watling Street and the Devil's Causeway. The former enters the county from York at Ebchester (Vindomora) and strikes for Riding Mill and thence to the Tyne at Corbridge (Corstopitum). From this place it passes due north to Hunnum, where it intersects the Roman Wall. Its general direction now is north-west *via* Colwell, Chesterhope, Risingham (Habitancum), and High Rochester (Bremenium), where it makes for a gap in the Cheviots. It passes over the ridge at Middle Golden Pot and crosses the border at Brown Hart Law (1664 feet), thus forsaking the low Rede valley in order to maintain its direction. The Devil's Causeway leaves the Roman Wall east of St Oswald's and runs straight to Watling Street at Bewclay, whence without deviation from its general north-east direction it passes for 16 miles to Whinny Hill. It then turns north past Longframlington to Whittingham, Hedgeley, and Wooperton, where it keeps to the Till valley for 21 miles. Its course here is absolutely straight till it arrives at Tweedmouth. On both sides of this road the fells are dotted with British camps, notably in the vicinity of Ingram. After the Roman occupation these roads were still utilised. New roads were constructed but they generally lay along the lowlands. These witnessed the tramp of armies with each war and period of border unrest but, in the quieter times of peace, afforded the only means of interchange of produce between town and country. Then, with the growth of the coal industry, waggon-ways were devised for drawing the coal along by primitive methods of haulage. There was, and still is, no system of canalisation, as the rivers have been so vastly improved as to meet practically all demands in this respect. With the advent of railways all was to change; the country was opened out so that even the remotest villages were brought within easy reach of the towns, but even here again we find a coincidence in general direction with valley and Roman Road.

Northumberland is served by the various lines of the North Eastern Railway and by a branch line of the North British Railway. The main line of the North Eastern enters the county from Gateshead by two routes, which have necessitated the construction of two magnificent bridges—the older High Level Bridge and the newer King Edward VII Bridge. It then proceeds due north, serving Morpeth, Alnmouth (the junction for Alnwick), Belford, Beal, and Tweedmouth. It crosses the Tweed by the Royal Border Bridge and reaches

**Newcastle-on-Tyne in the time of Queen Elizabeth**

(*Reconstructed from old records and drawings*)

**Newcastle-on-Tyne at the Present Day**

(*The bridge in the upper, and the swing bridge in the lower picture coincide with the site of the Roman Pons Aelii*)

Berwick. A branch line connects Norham, Coldstream, and Kelso. The branch line from Alnwick also passes to Cornhill near Coldstream, serving Wooler, Whittingham, and Eglingham. The fishing stations of Amble and North Sunderland are served by small lines. The towns along the Tyne Gap both east and west from Newcastle are also in the North Eastern network. A fine electrified system—a circular route—links up the towns on Tyneside east of Newcastle and the coast, returning *via* Backworth and Gosforth. The Avenue branch connects this line with Seghill, Seàton Delaval, Newsham, Blyth, Bedlington, Choppington, and Morpeth. The chief branch line to the west passes through the Tyne Gap to Carlisle, running *via* Corbridge, Hexham, Haltwhistle, and Gilsland, and forming a popular route for holiday-makers and tourists for the Roman Wall. The North Eastern Railway in the Tyne Gap has two branch lines, one from Hexham serving the Allen Dale and passing through Langley and Staward to Allendale town, the other serving Alston from Haltwhistle, using the South Tyne valley and passing through Featherstone, Coanwood, Lambley, and Slaggyford for Alston. In 1905 a branch was opened connecting Ponteland with South Gosforth.

The other important line is the North British, which, leaving Scotland, enters the county at Deadwater station at the head of the North Tyne valley. It passes through the country stations of Plashetts, Falstone, Tarset, Bellingham, Wark, Chollerton, and Chollerford, and joins the North Eastern Railway just below Warden.

Border Bridge, Berwick

Near Bellingham is the Reedsmouth junction, from which a branch line serves the Rede-Wansbeck valleys, passing through Woodburn, Knowesgate, and Scots Gap to Morpeth. From Scots Gap junction a branch leads to Rothbury *via* Longwitton, Ewesley, and Brinkburn.

The line from Newcastle to Berwick is the main line north and is traversed by the East Coast joint-stock expresses linking up the Scottish and English systems.

For many years one of the chief means of cheap travelling was along the Tyne from Tynemouth to Dunston by the steamers of the Tyne General Ferry Company, but for a variety of reasons these were stopped, and only on certain days in the summer months is it possible to make this delightful journey in the steamers chartered by the Tyne Trips Company. The Tyneside towns on the north bank are linked with those on the south by a system of direct ferries belonging to the Tyne Improvement Commission. These are at North Shields for South Shields, and at Howdon for Jarrow. Many plans have been put forward for joining these banks by tunnel and bridge, but without avail. A feature in the Tyne valley communications in the Newcastle area is the large number of fine bridges for pedestrians—notably the Swing Bridge, the High Level Bridge, and the Redheugh and Scotswood bridges. This then is, in epitome, Northumberland's system of communications, and it is remarkable to find that the principal routes of roads and railways and the sites of bridges keep in general alignment with those established by the Romans —that is to say, along the lines of least resistance.

## 24. Administration and Divisions.

The term Northumberland in its contracted modern sense dates from 1065. It is mentioned in the Anglo-Saxon Chronicle relative to the rebellion in the North. Mention is, however, omitted from the Domesday Book. In 1131, when Odardus was made the first Sheriff, an account of the taxes of the county appeared in the Great Roll of the Exchequer. In the early Saxon times administration was at first of a purely family nature; then it grew to be tribal, and as expansion progressed with the formation of the shares or shires, or with the coming of the Normans who called the residual areas counties, it was necessary to have some definite local governing body. In a similar manner has administration developed to the present day, when each new advance in legislation complicates the labour of both elected administrators and paid officials.

In Saxon times the *ealdormen* were virtually the governors of the shires. They relegated their duties to the sheriffs, who held sway alike over common freedmen and labourers. With but slight alteration the Normans perpetuated the Saxon system. In the reign of Henry I the first Norman Sheriff was appointed and the Barony of Prudhoe was given to the Umfravilles, who also held the franchise of Redesdale. In the reign of Edward I Northumberland included a district extending from the Tees to the Tweed and contained many special liberties, namely Durham, Norham, and the wapentake of Sadberge, which were subject to the Bishops of Durham. Further, the Archbishop of York held the liberty of Hexham, the King of Scots held Tynedale, the Earl of Lancaster held Emildon, and Umfraville, Earl of Angus, possessed the liberty of Redesdale. This complicated matters of administration, for they were all free from the general jurisdiction of the county. For certain offences first the lordship of Tynedale was included in the county (1495) and then the liberty of Hexham was taken away in 1572, but the detached portions of the County Palatine, Norhamshire, Bedlingtonshire, and Islandshire, remained under the Bishop Princes till 1844, when they also were incorporated in the county. For purposes of local government Northumberland was split up into *hundreds* or *wards*. They existed as far back as 1295, for in the Hundred Roll of that year we find the wards of Bamburgh, Coquetdale, Glendale, and Tindale recorded. It is now divided into nine wards, Bamburgh, Bedlingtonshire, Castle, Coquetdale, Glendale, Islandshire, Morpeth, Norhamshire, and Tynedale. Still further subdivision resulted in parishes, of which there are 511 civil and 193 ecclesiastical in Northumberland. With the growth of towns in the more populous areas it soon became apparent that they required special treatment, so that we find Newcastle is made a county by itself in the reign of Henry IV (1400) and possessed special maritime jurisdiction. The shire court for Northumberland had been held at different times at Alnwick, Morpeth, and Newcastle. By a Statute of 1549 this court was ordered to meet in future at Alnwick Castle. The Sheriffs of the county at that time, who were also "Wardens of the

Border Marches," had been wont to appropriate the taxes raised in the shire. Thenceforward they were compelled to present accounts to the Exchequer. The assizes were held in Newcastle, and on the approach of the itinerant justices to the county either from the south *via* York, or west *via* Carlisle, they were met either at Gateshead or Fourstones by the Prior of Tynemouth, the Bishop of Durham, the Archbishop of York, and the King of the Scots. They pleaded their liberties and were exempted from the King's writ, and performed their own offices of sheriff and coroner.

Although the Great Charter was passed in 1215, the county, as such, does not appear to have had much voice in national affairs. In 1272, in the reign of Edward I, two knights of the shire certainly appear in local registers, and in the first complete Parliament (1294) Bamburgh, Corbridge, and Newcastle each had two members. From 1297 Newcastle alone was represented, until 1524, when Berwick sent two members; in 1553 Morpeth also sent two. In 1832, at the time of the Reform Bill, the county was represented by only four members in two divisions. In 1885 there were four members in four divisions, and to-day the county is represented by seven members.

Having passed in brief review this gradual evolution of our county's administration, let us now examine it as it is to-day. At the head of affairs stands the Lord Lieutenant of the County, and Custos Rotulorum, the Duke of Northumberland, who represents the Crown. Then under him is the High Sheriff, whose duties, whilst still resembling those of the old Norman sheriff, are mainly legal. He is responsible for the peace of the county and attends the judges when on circuit. The major portion of the administrative work devolves upon the County Council. Its number, as fixed by the Local Government Board, is constituted of 60 councillors elected in rotation for three years, and 18 county aldermen elected for six years. These with the Chairman and Vice-chairman bring the number up to 80. It has a multitude of duties connected with education, the upkeep of bridges and roads, the care of lunatics and public buildings, the prevention of contagious diseases amongst animals, and the appointment of coroners. The county is further administered by Rural District and Urban District Councils, of which there are respectively 10 and 21. In other areas where the population is over 300 a Parish Council must be elected, and it can administer its own Poor-rate.

The rateable value of the county is £5,660,940[1], the rental £2,268,169, and there are 11,381 paupers.

The county is in the north-east circuit, the Assizes being held at Newcastle at the Moot Hall. There are also Quarter Sessions held at Newcastle, and Petty Sessions at Alnwick, Hexham and Morpeth, each with its own Justices of the Peace.

The county is in the diocese of Newcastle in the Province of York, and the See is of recent creation, dating only from 1882. Prior to that, it was in the diocese of Durham (1291) and formed the Archdeaconry

[1] £2,398,964 excluding the County Boroughs of Newcastle and Tynemouth.

of Northumberland, with the deaconries of Alnwick, Bellingham, Bamburgh, Corbridge, and Morpeth.

In 1845 the Archdeaconry of Lindisfarne was formed and was subdivided into the rural deaneries of Alnwick, Bamburgh, Morpeth, and Rothbury; the Archdeaconry of Northumberland then including Bellingham, Corbridge, Hexham, and Newcastle. The creation of the see did not alter these deaneries but only added one or two more.

## 25. The Roll of Honour of the County.

Lying close upon the borders, this county in the early days was always in a condition of turmoil. What with border feuds, punitive expeditions, and counter pillagings, its stern and rigorous climate, and grimly rugged landscapes, it was no home for the weakling. But these hardening factors of climate and environment acting on a good racial stock have resulted in the Northumbrian, and it is with just pride that he can contemplate the Roll of Honour of his county. Out of the scores of famous men, many of world-wide repute, we can but mention a few, and only those who have attained a wide renown.

At sea, Northumberland gained celebrity through the exploits of Sir Chaloner Ogle, Robert Roddam, and Cuthbert, Lord Collingwood. Ogle of Kirkley (1680–1750) became an Admiral, as did Roddam of Little Houghton (1720–1808), but it is the name of Collingwood, the hero of Trafalgar, eulogised by Nelson, that chiefly attracts us. Cuthbert, Lord Collingwood (1750–1810), born at Newcastle, was a school-fellow of John Scott, afterwards Lord Eldon. He entered the Navy at the age of eleven, and in 1805 was second in command at Trafalgar. For his signal services to the nation he

**Lord Collingwood**

was created Baron Caldbourne and Hethpole in Northumberland, and subsequently became Commander-in-Chief in the Mediterranean. He died in Port Mahon and was accorded a hero's tomb beside Nelson in Westminster Abbey. The Collingwood monument at

Tynemouth, another in St Paul's, and another in St Nicholas' Cathedral, Newcastle, testify to his greatness.

It is in the engineering world, however, that Northumberland is perhaps most famous. The Chapmans, father and son, were both distinguished, one being renowned at the Royal Society (1758), while the son (1750–1832) is called to mind by every skew-arched

George Stephenson's Engine

bridge in the country. John Blenkinsop of Walker (1782–1831), the pioneer in steam locomotion, tried his first locomotive at Coxlodge in 1812, with George Stephenson a critical spectator. George Stephenson, father of railways (1781–1848), was born at Wylam. He became an engine-wright and an inventor, and in 1814 his first locomotive actually drew eight loaded carriages up an incline of 1 in 450 at five miles per hour.

He also invented the safety-lamp contemporaneously with Davy. Newcastle honours his memory with a handsome monument. His son, Robert Stephenson (1803–1859), who was born at Willington, "the Builder of Bridges," was not less distinguished, and was honoured

**Earl Grey**

by burial in Westminster Abbey. The last of the galaxy of mechanical experts who strove so much to add to man's knowledge and to alter the conditions of human life was William, Lord Armstrong (1810–1900). His life has been aptly summarised in four words—hydraulics, artillery, Elswick and Cragside. At Elswick he made his renowned hydraulic cranes and Armstrong guns, and established one of the greatest factories in the world.

In Parliamentary affairs three figures stand out in bold relief. These are Lord Eldon, Lord Stowell, and Earl Grey. William Scott, Lord Stowell (1745–1836), of Heworth, was a Judge of the High Court of Admiralty. His younger and more brilliant brother, John Scott, Lord Eldon (1751–1838), who was born at Newcastle, ran his career abreast of William, becoming Attorney-General in 1793 and Lord Chancellor in 1801. Greatest in Northumberland's political history is Earl Grey (1764–1845) of Falloden. He had the mind of the reformer and foreshadowed many of the political difficulties of to-day. As Mr Grey, he was First Lord of the Admiralty, and later became Secretary for Foreign Affairs and Leader of the House of Commons. He passed the Great Reform Bill and the Bill for the Abolition of the Slave Trade. He was immensely popular, and Grey's Monument and Grey Street, Newcastle, recall his memory.

In the realm of natural history Northumberland has contributed not a little to the sum of our knowledge by the work of an enthusiastic group of naturalists, Joshua Alder (1792–1867) of Newcastle; George Clayton Atkinson (1808–1877) also of Newcastle, who founded the Natural History Society; W. C. Hewitson, the ornithologist and oologist; Albany Hancock (1806–1873) and John Hancock (1808–1890), two brilliant brothers, the latter an ardent ornithologist, a younger brother of Albany, and a worker rather than a writer. Albany

Hancock was honoured with a gold medal of the Royal Society, and it is to the memory of the two brothers that the Museum in Newcastle stands. But Thomas Belt (1832–1878), the author of *The Naturalist in Nicaragua*, one of the best books of scientific travel ever written, has a wider fame probably than any of the foregoing.

History has had its distinguished representatives in our county, and the name of John Collingwood Bruce (1805–1892), the schoolmaster antiquary of Newcastle, is synonymous with most that is worth knowing of Roman History in Britain. Associated with his name is that of John Clayton (1792–1890), the antiquary of Newcastle, the owner of the Roman Wall for miles, the inspirer of antiquarian research and, along with Richard Grainger (1797–1861), the maker of modern Newcastle. George Tate (1805–1869) must also be recorded as a noted Alnwick historian.

Nicholas Ridley, Bishop of London, Northumberland's greatest prelate, was born at either Willimoteswick or Unthank Hall, about 1500, and died at the stake with Latimer in 1555. Robert Rhodes was the originator of the famous lantern tower of St Nicholas, Newcastle. He sat as member in 1427 and died in 1474.

Northumberland boasts of few poets, one of the few who has gained other than local distinction being Mark Akenside (1721–1770), who was both poet and doctor. He wrote *The Pleasures of Imagination* in 1743, a prolix and somewhat dull poem which is now read by few. She is stronger in the sciences, as might be imagined, and has turned out more than one distinguished mathematician.

Charles Hutton (1737–1823) of Newcastle, became Professor of Mathematics at the Royal Military Academy of Woolwich and was also distinguished at the Royal Society, and Sir George Airy (1801–1892), born at Alnwick, after a distinguished career at Cambridge, where he was Senior Wrangler, became Astronomer Royal in 1836.

Ralph Gardiner (born 1625) of Newcastle, strove against the practice of monopoly. He resided at Chirton, North Shields. Practically no inhabitants of Tyneside could "bake, brew, steep malt, or build ships" unless they bought the freedom of Newcastle. This monopoly was granted by Royal Charter, and for striving against it Gardiner suffered imprisonment and died in obscurity later than 1682.

It is not a little remarkable that, far from the great art centres, and amid the uninspiring surroundings of a vast manufacturing town, a school of artists should have arisen in Newcastle scarcely inferior to those of Norwich and Bristol. Few places in England can show a longer list of painters and engravers. Ralph Beilby did not, perhaps, attain much fame as an engraver, but his name must find record as the discoverer and master of Thomas Bewick (1753–1828), one of the greatest of English wood-engravers, whose work in his *History of British Birds* and the companion *Quadrupeds* has in its way never been surpassed. His brother, John Bewick (1760–1795), who was also born at Cherryburn near Ovingham, was his pupil, and though far inferior to him as an engraver, was a good draughtsman. Two other pupils of Thomas Bewick, Robert Johnson (1770–1796), born at Shotley, and Luke Clennell of Ulgham (1781–1840), attained considerable distinction as watercolourists—the latter especially being a pioneer in this branch of art. "His rustic groups," says Redgrave, "were admirable,

Thomas Bewick

full of character and nature." E. Landells (1808–1860) was yet another pupil of Bewick, and was one of the little group connected with the founding of *Punch*.

John Martin's canvases, crowded with myriads of figures, were familiar to all in the first part of the nineteenth century. Born near Hexham in 1789, he was the pupil of Muss of Newcastle, enamel-painter to George IV, and died in 1854. The watercolourist, George Balmer, born at North Shields at the commencement of the nineteenth century, is widely known by his contributions to Finden's *Ports and Harbours*, and among lesser lights in the school of watercolour painting J. H. Mole (1814–1886) of Alnwick, and James Peel of Newcastle (1811–1906) must be mentioned. James Wilson Carmichael (1800–1868), a native of Newcastle who at one time worked with Balmer, was among the best of our English marine painters, working both in oil and watercolour.

In Myles Birket Foster we have an artist whose name is perhaps as familiar as that of any other watercolourist alive or dead. He was born at North Shields in 1825 and lived to see himself famous, while at the present day his pictures continue to appreciate in value. Of very high merit too was William Nicholson (1784–1844), yet another Novocastrian, who was one of the founders of the Royal Scottish Academy. Finally we have the Richardsons, father and son, both named Thomas Miles (1784–1848, and 1813–1890). The father founded the Newcastle Water-colour Society in 1831, and the charming Italian landscapes of the son are greatly admired by connoisseurs. John Graham Lough, the sculptor (1798–1876), is best known by the statue of Queen Victoria in the Royal Exchange. He also made that of George Stephenson at Newcastle and of Collingwood at the Tyne entrance. In the realm of art our county—and Newcastle especially—has every reason to be proud.

Perhaps at the head rather than at the foot of our Roll of Honour should come Grace Darling (1815–

**Grace Darling**

1842) of Bamburgh, the heroine of the Longstone, whose boat, in which her gallant deed was performed, now reposes in the Dove Marine Laboratory at Cullercoats.

## 26. THE CHIEF TOWNS AND VILLAGES OF NORTHUMBERLAND.

(The figures in brackets give the population in 1911. The figures at the end of each section are references to the pages in the text.)

**Allendale Town** (2185), situated at an elevation of 800 feet, on the east bank of the Allen, 10 miles south-west from Hexham, is a popular summer resort. It was formerly important as a centre of lead-mining. The principal mines were at Coalcleugh and Allenheads, but they were closed in 1910. The township was formerly divided into seven divisions called grieveships. (pp. 14, 66, 68, 84, 149.)

**Alnmouth** (542) is a pretty fishing-village five miles south-east from Alnwick. It forms a good holiday centre for places of historic interest, notably Alnwick, Warkworth, and Dunstanburgh, which are within easy reach. It is claimed that a great Synod was held here in 684 which chose Cuthbert as Bishop of Lindisfarne, and that Alnmouth is the Twyford of Bede. (pp. 8, 20, 46, 91, 145.)

**Alnwick** (7761), the county town, is also a market town, and stands on the south bank of the Aln, five miles from the sea and 34 miles north by west from Newcastle. This is the historic home of the Percys. In the centre of fine fertile lands and delightful scenery, it presents the appearance of a sleepy, old-world town. Its quaint bars, gates, and pants, its toll-booth, and pound or pinfold all testify to its antiquity. Alnwick is distinctly agricultural and has markets for dairy-produce, cattle, horse, and sheep fairs, and lamb, wool, and hiring fairs. The castle stands on the south bank of the Aln. Before its walls Malcolm, King of Scotland, fell in 1093, and in 1174 William the Lion was taken prisoner. It has been the resting-place of many English monarchs. The Hotspur Bar testifies to the memory of that gallant fighter, and the Tenantry Column is another interesting landmark. The neighbouring Hulne Abbey dates from 1250, and was the home of Carmelite monks. The church of St Mary and St Michael is a fine example of the Perpendicular style. (pp. 15, 20, 29, 98, 100, 103, 128, 134, 135, 137, 139, 145, 147, 151, 153, 154, 160, 162.)

**Amble** (4881) is a seaport on the Coquet estuary, one and a half miles south-east from Warkworth. Collieries have contributed to its prosperity, and the harbour has been deepened to accommodate vessels of 2000 tons. (pp. 11, 20, 45, 46, 80, 91, 103, 104, 130, 147.)

**Annitsford.** *See* Weetslade. (p. 83.)

**Ashington-with-Hirst** (24,583), an urban district, six miles east from Morpeth, owes its phenomenal growth to the development of collieries. Hirst numbers 16,428 people. (p. 83.)

**Backworth** (2235), near Earsdon, is a colliery village and railway junction, situated on the Benton Curve of the N.E.R. (pp. 67, 83, 147.)

**Bamburgh** (433), a popular resort of great antiquity and rich associations, with a fine castle, is situated on the coast, three and a quarter miles north-east from Lucker Station. The history of Bamburgh is the history of its castle. It caps the Whin Sill, which is 75 feet thick at this point. Its restoration was undertaken by Lord Armstrong, who did not live to see it completed. At the head of the village stands St Aidan's Church (early thirteenth century), and in its churchyard is the grave of Grace Darling. Bamburgh is the centre for the Farnes. (pp. 9, 15, 29, 31, 49, 66, 95, 98, 131, 133, 151, 152, 163.)

**Bedlington** (25,440), five miles south-east from Morpeth, stands in an elevated position near the river Blyth, where this river gives true "dene" scenery. The name "Bedlingtonshire" speaks of its history. Bishop Cutheard purchased it about the tenth century and gave it to the see of Durham. It thus was subject to the Bishop Princes in civil and ecclesiastical matters till 1535, when its peculiar privileges were taken away. The church of St Cuthbert is built on the site of a Saxon church, near where the relics of St Cuthbert rested in 1069. (pp. 3, 21, 66, 147, 151.)

**Bellingham** (1358), a small market town with a fine situation on the north bank of the North Tyne, 17 miles north-west from Hexham, stands at the entrance to the wildest parts of Northumberland. The church of St Cuthbert was founded in the eleventh century and is typically Early Norman. Bellingham is well known for its fair and sports. It was the site of an iron industry. It has wool fairs and auction marts for cattle and lambs, and witnesses annually the ancient custom of "Riding the Fair." Hareshaw Linn is a fine waterfall quite near. (pp. 83, 103, 147, 154.)

**Berwick-upon-Tweed** (8110), a county in itself, is a seaport and market town situated 64 miles north by west from Newcastle on the north bank of the river Tweed. It is of great antiquity, as its name testifies, and figures largely in British history, alternating between the Scots and English. It is a walled town with the remains of an old castle. The walls form the chief promenade of the town and date from Elizabeth's reign. It is approached by the Royal Border Bridge, designed by Robert Stephenson and opened by Queen Victoria in 1850. This forms the main entry into Scotland from the south. The town is sombre looking, and is singularly devoid of architectural beauty or historic buildings. Its principal imports are wood and grain. Manure and fish—notably salmon—are exported. It has fish and cattle markets, cattle fairs, and hirings. (pp. 6, 29, 49, 51, 52, 53, 65, 68, 80, 88, 89, 91, 138, 147, 148, 152.)

**Blanchland** and **Shotley** (705), situated 12 miles south of Hexham in lovely moorland on the Derwent, are immortalised by Besant in *Dorothy Forster*. An abbey was founded here in 1174 by the White Canons. (pp. 8, 14, 68, 128.)

Corbridge

**Blyth** (28,490) lies on the south bank of the river Blyth, nine miles north from North Shields. It has a fine harbour with piers, and possesses fine staithes and spouts for loading colliers. It has a huge and still increasing export of both coal and fish, and is the centre for the output of the collieries, of which there are no less than forty in the vicinity. It possesses graving and dry docks, and shipbuilding and repairing yards. Fishing (herring and salmon) is at its height in July and August. (pp. 21, 24, 29, 37, 44, 53, 80, 81, 82, 83, 88, 91, 147.)

**Corbridge** (2213), 16 miles to the west of Newcastle, occupying a fine position in the Tyne valley, is a village of great antiquity and is on this account popular as a resort. It is celebrated for its Roman remains of *Hunnum* and *Corstopitum*, the latter at present being excavated and yielding fine remains. It stands on the Watling Street and must have been a very important Roman station. The church of St Andrew, built on the site of a monastery (771), was probably founded by St Wilfrid and contains some fine Saxon and Roman remains. The river is spanned by a bridge, the only one which survived the great Tyne floods of 1771. (pp. 18, 29, 103, 108, 111, 118, 119, 120, 121, 123, 144, 147, 152, 154, 167.)

**Cramlington** with **Shankhouse** (6376), 10 miles north-east from Newcastle, has extensive steam-coal collieries. (p. 84.)

**Cullercoats** (in Tynemouth), a quaint fishing village, beloved of artists, on a fine bay, one and a half miles from Tynemouth, formerly exported salt and coal. The dress of the fisherwomen is a familiar sight on Tyneside. Here is established one of the most modern buildings in the country for the prosecution of researches in fisheries, the Dove Marine Laboratory, which is incorporated with Armstrong College and owes its inception to the late W. Hudleston, F.R.S. (pp. 24, 26, 29, 30, 37, 40, 41, 46, 53, 91, 93, 142, 163.)

**Dudley**. *See* Weetslade.

**Earsdon** (3777), four miles north-west by north from Shields, possesses very extensive collieries in its parish. Here the victims of the terrible Hartley disaster of 1862 are buried. (p. 66.)

**Haltwhistle** (3979), a small market town, 36 miles west from Newcastle, stands on the north side of the South Tyne river on the N.E.R. Fairs and hirings take place here, and it is a great centre for tourists visiting the Roman Wall. Its recent development is due to its collieries. (pp. 14, 17, 18, 94, 122, 147, 149.)

**Hartley** with **Seaton Sluice** (1688), a township three miles from Whitley Bay, is engaged in mining and fishing. St Mary's Island can be reached from here. (pp. 8, 30, 37, 40, 43, 67, 82, 89.)

Frið Stool, Hexham Abbey

**Haydon Bridge** (2297), a village lying on both banks of the South Tyne, 28½ miles west from Newcastle, is a centre for visitors to the Roman Wall. (pp. 18, 103.)

**Hexham** (8417), an ancient market town, is situated upon a flat rich terrace of the Tyne, just below the confluence of the North and South Tynes, 20 miles west from Newcastle. It is the junction of the North Eastern and North British Railways. Its antiquity is great, and this is shown by the quaint streets and their names, the Keep, Abbey, and Moot Hall. In church history it occupies a pre-eminent position, and the Abbey has been spoken of as a "text book of Early English architecture." Historically, Hexham is connected with the Wars of the Roses. It possessed a famous tanning industry, and used to produce large quantities of tan gloves (Hexham tans). It is now the centre of numerous market gardens and nurseries. (pp. 17, 84, 100, 120, 123, 125, 127, 128, 130, 147, 149, 151, 153, 154, 162, 169.)

**Holy Island** (359), or Lindisfarne, is reached from Bamburgh *via* Budle, a distance of nearly five miles. The approach to it is celebrated in Scott's *Marmion*. It consists only of one street of fishermen's houses, but owing to its interesting history is a great holiday resort. (pp. 3, 9, 10, 50, 51, 91, 93, 98, 123, 125, 126, 130, 151.)

**Killingworth** (in Longbenton), six miles north by east from Newcastle, is a colliery village, mainly noteworthy in connection with George Stephenson, who was a brakesman at the colliery, where he made his first improvements in steam engines. (pp. 29, 67, 83.)

**Longbenton** (12,443), a large village four miles north-east by north from Newcastle, has a mining population, which includes Forest Hall, Killingworth, and West Moor.

**Morpeth** (7433) is a very old market town in a richly-wooded agricultural district, 15 miles from Newcastle, on the Wansbeck, but does not figure largely in the history of the county. The arms were granted to the town in 1552 by Edward VI. The parish church of St Mary dates from the fourteenth century. Morpeth has many quaint hostelries, particularly "The Queen's Head," where John Scott (Lord Eldon) stayed after eloping with Bessie Surtees. Newminster Abbey is quite near; it was founded by Ranulph de Merlay in 1139, and, owing to its proximity to the Great North Road, was often visited by royalty. (pp. 11, 20, 29, 44, 100, 130, 145, 147, 151, 152, 153, 154.)

**Newbiggin** (3466), a popular watering-place and fishing-village, stands on a pretty bay between two headlands, nine miles east of Morpeth. It used to be a great corn-exporting port, though it has now decayed. The conspicuous landmark of St Bartholomew's Church stands on Newbiggin Point. Newbiggin has large fisheries and is in the vicinity of huge collieries. (pp. 8, 44, 45, 46, 67, 91.)

**Newburn** (4260), five miles west of Newcastle-on-Tyne, owes its ancient site to the then fordable nature of the Tyne. The church of St Michael and All Angels was built in Saxon and Early Norman times. In 1346 David of Scotland passed south to Neville's Cross, and here in 1640 the disastrous rout took place when Leslie defeated Lord Conway. George Stephenson was educated and married here, and it is the seat of Spencer's Steel Works. (p. 81.)

**Newcastle** (250,825) is a city and county of itself, lying 10 miles west of Tynemouth on the north bank of the river Tyne. It is typical of the growth of towns at the intersection of two "lines of weakness," the Great North Road and the Tyne. This, the largest and most important town in the county, dates from the time of Hadrian, who built Pons Aelii in 120 A.D. In Saxon times Newcastle was known as Monkchester. In 1080 it was fortified by Robert Curthose, and later a "new castle" was erected by William Rufus and finally finished by Henry II in 1177. Newcastle was created a county by Royal Charter in 1400. In 1838 the Newcastle and Carlisle Railway was opened, and in 1850 the Newcastle Central Station and that magnificent example of scientific engineering—the High Level Bridge—were inaugurated by Queen Victoria. The fine streets are the results of the joint work of Richard Grainger and John Clayton. On approaching the town from the south, the main objects are the magnificent bridges, the Castle, and St Nicholas' spire. Newcastle has very large markets, and is the seat of the Assizes and Quarter Sessions, which are held in the Moot Hall. With regard to open spaces

**Newcastle Cathedral**

it is singularly fortunate, possessing the magnificent gifts of Lord Armstrong to the city—the Armstrong Park and Jesmond Dene, in addition to 161 acres of park land, and some 927 acres of land called the Town Moor over which the Freemen of Newcastle possess certain rights. Many of the churches are celebrated. St Nicholas' Church was raised to the dignity of a cathedral in 1882, in the Province of York. St Andrew's Church was built by King David of Scotland, and St John's dates from the time of Edward I. The ruins of the town walls, the castle, and the names of many of the streets, such as High and Low Friar Street, Pilgrim Street, The Close, The Sandhill, etc. all indicate its antiquity. The Museum of Natural History is in Barras Bridge. The fine

**Armstrong College, Newcastle**

buildings and laboratories of Armstrong College and the Medical College in the University of Durham form the centre of education on Tyneside. The Lough Models are carefully housed in the Elswick Park. The port of Newcastle has been linked with the Gateshead side by the High Level Bridge (built in 1850, at a cost of £491,000), the Swing Bridge, which occupies the site of the Pons Aelii and is worked by hydraulic power (1876), the Scotswood Bridge (1831), and the Redheugh Bridge (1867). The coal industry originated in 1350, when the burgesses were

North Shields

granted a charter to dig for coal. "Elswick" was established in 1847. The Newcastle Infirmary fronts the Armstrong College and was opened by King Edward VII in 1906, when he also opened Armstrong College and the Edward VII Bridge. (pp. 14, 22, 59, 60, 73, 75, 77, 78, 79, 81, 84, 98, 99, 100, 102, 108, 121, 123, 125, 128, 130, 131, 138, 144, 146, 147, 149, 152, 153, 154, 155, 157, 158, 159, 160, 162, 171, 172.)

**New Delaval** (2196), a quarter of a mile north-west from Newsham, has collieries and brickworks. (p. 43.)

**North Shields** (*see* Tynemouth) is a seaport in the county borough of Tynemouth, standing on the northern bank of the river Tyne and about eight miles east-north-east from Newcastle. As a port its growth is of the last two centuries, though it was a settlement as early as the reign of Edward I. It has important docks, quays, and works in connection with a very thriving fishing industry. The "High Lights" and "Low Lights" lighthouses are conspicuous features, and the old "Wellesley" training ship is reminiscent of earlier days. Fuller details have been given under the chapter on Shipping and Trade. (pp. 38, 40, 44, 53, 54, 68, 74, 78, 82, 89, 91, 92, 93, 149, 162.)

**Prudhoe** (4734), on the south bank of the river Tyne, is 10½ miles west of Newcastle in the Hexham Division. The fine castle dates from Henry II's reign, being built by Odinel de Umfraville. It was transferred in 1381 to Henry Percy, first Earl of Northumberland, by marriage. (pp. 137, 150.)

**Rothbury** (1147), near the Coquet, is a township and head of a county court district in the Hexham Division. Its fine air, scenery, and fishing attract numerous visitors. In feudal times there were four gallows in Coquetdale, one being situated near Gallowfield Braes. There are numerous traces of an early British occupation in this district, the environing fells being the site of many camps. Cragside bears testimony to the greatness of the late Lord Armstrong, who transformed a veritable wilderness into one of the finest seats in the county. Numerous relics of glaciation are seen in the fells. (pp. 12, 13, 20, 23, 29, 31, 140, 141, 147, 154.)

**Seaton Delaval** (5288), part of the parish of Delaval, which includes part of the townships of Seaton Delaval and the villages of Hartley, Seaton Sluice, New Hartley, and New Delaval, is seven miles north of North Shields. Its parish church—the Church of Our Lady—the gift of Lord Hastings, was built by Hubert de la Val in the eleventh century, and is all that remains of the ancient castle. In 1786 the township was chosen to confer the title of Baron upon Sir John Hussey Delaval. Seaton Delaval Hall, a ruin, has a fine Doric portico. Seaton Delaval is a colliery village. (pp. 67, 123.)

**Seghill** (2049) is a colliery village on the North Eastern Railway, six and a half miles north-west of North Shields. (pp. 66, 83, 84, 103, 147.)

**Stannington** (1194) is a great centre for cyclists on the North Road, 10 miles north from Newcastle. It is in the Stannington Vale. Near it is the training school of the Netherton Reformatory. (p. 21.)

**Tweedmouth** and **Spittal** (4965) form a township in the Borough of Berwick. It is about a quarter of a mile from Berwick at the south end of the bridge, and is a port engaged in the salmon industry. Its dock will take ships of 500 tons. It imports timber and manufactures spades, shovels, and agricultural implements. It is partly engaged in boat-building and herring-curing and has manure works. (p. 145.)

**Tynemouth** (58,816) has a fine situation on the promontory on the north bank of the Tyne entrance. The Tyne is here protected by magnificent stone piers. The North Pier, commenced in 1854, was not completed till 1893, and was partly destroyed in 1897, but has since been rebuilt. A fine lighthouse is situated at the end of this promenade. Quite near are the Black Middens, with the Lough Monument of Lord Collingwood overlooking them. Its famous abbey stands in the castle grounds, on the site of a church founded in 627 A.D. by Edwin of Northumbria and rebuilt by Oswald. It was repeatedly attacked by Danes and eventually destroyed. The Benedictine Priory was

Tynemouth

founded by Waltheof, Earl of Northumberland, in the reign of William I. Malcolm Canmore and his son Edward were buried here in 1093. (pp. 7, 8, 9, 15, 22, 24, 26, 30, 31, 40, 41, 42, 43, 53, 54, 98, 100, 125, 128, 129, 130, 138, 149, 152, 156, 175.)

**Wallsend** (41,461), four miles east-north-east from Newcastle, is a municipal borough standing on the Tyne and served by two stations on the N.E.R. Here is the site of *Segedunum*, the eastern extremity of the Roman Wall. The coal mines are renowned all over England. It is the centre for all the great shipbuilding yards of central Tyneside, and it was here that the huge liner *Mauretania* was designed and constructed. (pp. 79, 80, 81, 83, 105, 108, 121, 144.)

**Warkworth** (710), on the Coquet, seven and a half miles south-east of Alnwick and half a mile from the sea, is a holiday resort in fine wooded scenery, famed for its castle and hermitage. Its port is Amble. The church of St Lawrence dates from the eighth century. The castle stands on a hill on the south bank of the Coquet, and was built by Roger Fitz Richard, first Lord of Warkworth, in 1158. The Hermitage (about 1400) is half a mile from the castle on the south bank, amidst lovely wooded scenery. It is hewn out of a sandstone cliff. (pp. 82, 100, 133, 134.)

**Weetslade** (6701), a civil parish with a station called Annitsford in Dudley village, is seven miles north by east from Newcastle. Dudley is a colliery village, as is Annitsford.

**Whitley Bay** (11,463), two and a half miles north-east from North Shields on the coast, is a popular holiday resort with fine links. The St Mary's Lighthouse (1898) is a prominent landmark. (pp. 8, 40, 41, 43, 52, 53.)

**Willington Quay** (in Wallsend), on the Tyne, four and a half miles east by north from Newcastle, on the N.E.R., is noted as the birthplace of Robert Stephenson. It has large rope-works. (pp. 81, 157.)

**Wooler** (1336) is a small cattle-market town surrounded by picturesque scenery, six miles from the base of the Cheviot range. Chillingham Castle is within easy reach. (pp. 11, 15, 19, 29, 68, 100, 139, 147.)

**Wylam** (1312) is a village on the north bank of the Tyne, nine miles west of Newcastle. Near here George Stephenson was born. The Manor of Wylam was formerly attached to the monastery at Tynemouth. (pp. 7, 8, 84, 143, 156.)

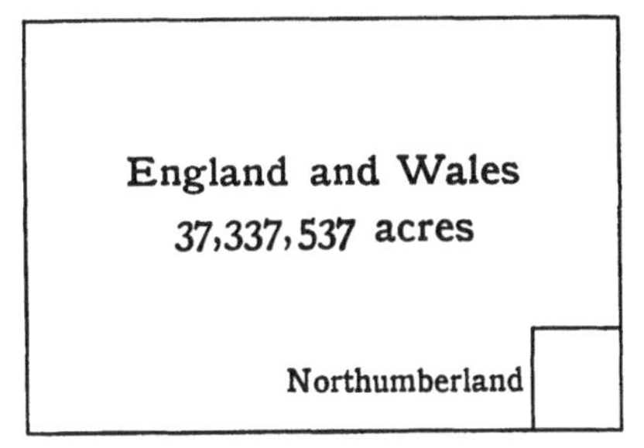


Fig. 1. Area of Northumberland (1,291,515 acres) compared with that of England and Wales

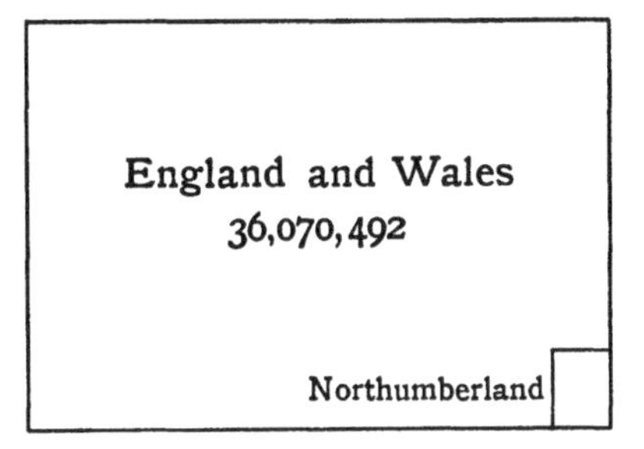


Fig. 2. Population of Northumberland (696,893) compared with that of England and Wales in 1911

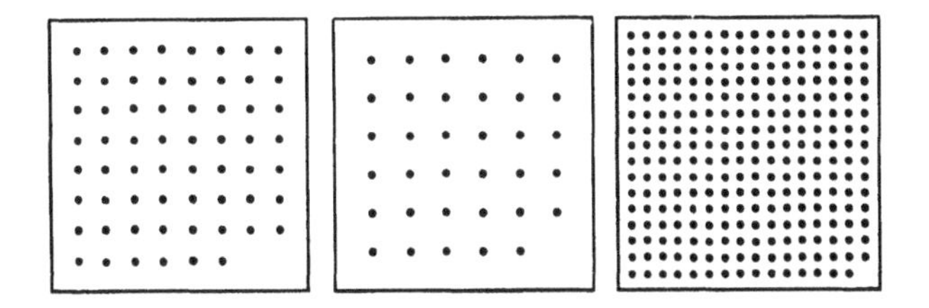

England and Wales 618 Northumberland 346 Lancashire 2554

Fig. 3. Comparative Density of Population to the Square Mile in 1911

(*Each dot represents ten persons*)

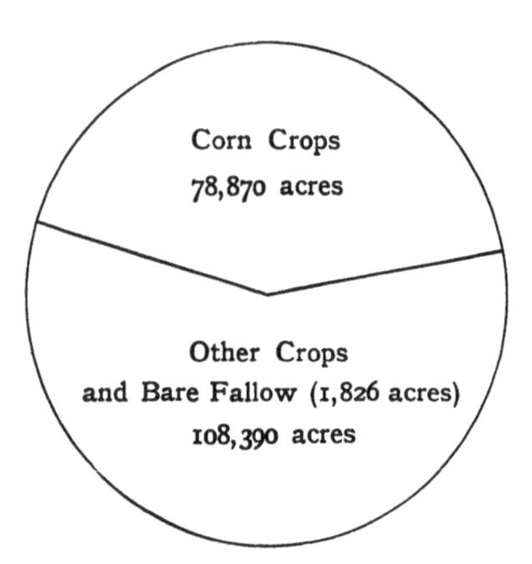


Fig. 4. Proportionate Area under Corn Crops compared with that of other land in Northumberland in 1912

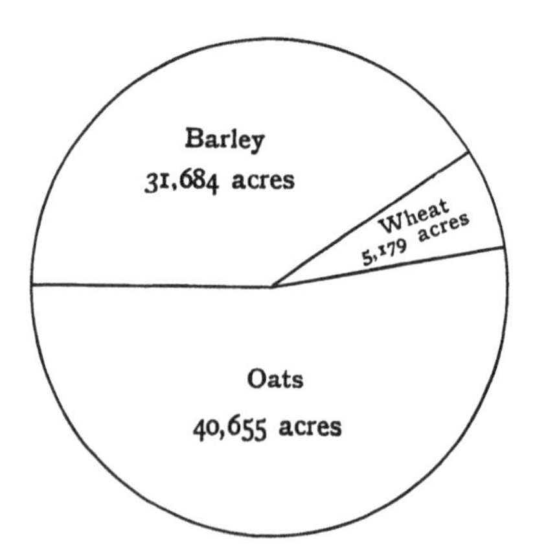


Fig. 5. Proportionate Area of chief Cereals in Northumberland in 1912

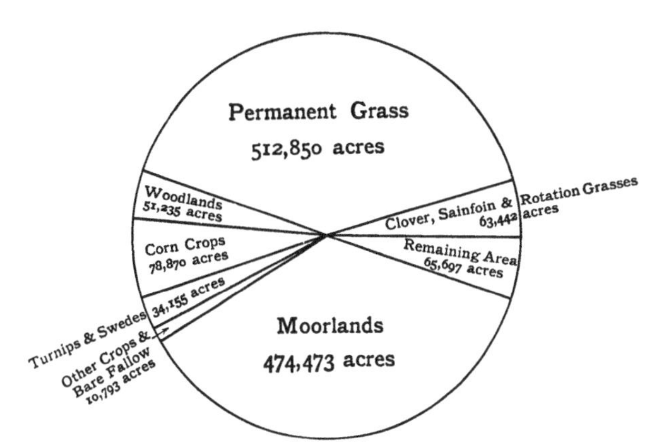


Fig. 6. Proportion of Permanent Grass to other Areas in Northumberland in 1912

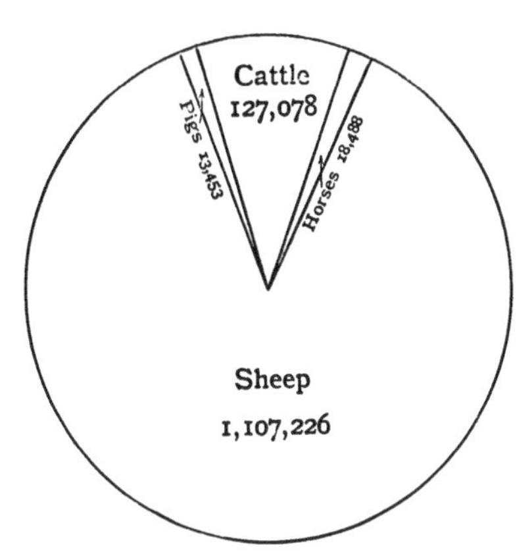


Fig. 7. Proportionate numbers of Live Stock in Northumberland in 1912

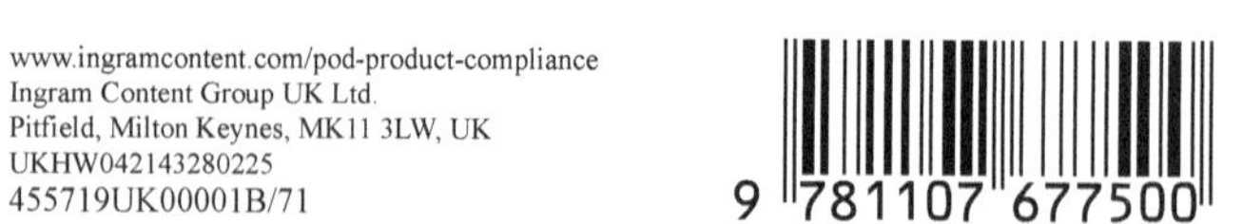


www.ingramcontent.com/pod-product-compliance
Ingram Content Group UK Ltd.
Pitfield, Milton Keynes, MK11 3LW, UK
UKHW042143280225
455719UK00001B/71